# The Thomas-Morse MB-3

## America's First Indigenous Fighter

**Jan Forsgren**

**Acknowledgements**
Simon Beck, Dan Hagedorn, Tor Johnsson, Chris Stanton

A special Thank You goes to Colin Owers, without whose considerable assistance this book could not have been written. Thank You also to Greg VanWyngarden for photos.

**Great War Aviation Centennial Series #60**

**Acknowledgements**
Color aircraft profiles © Bob Pearson. Purchase his CD of WWI aircraft profiles for $50 US/Canadian, 40€, or £30, airmail postage included, via Paypal to Bob at: **bpearson@kaien.net**

For our aviation books, please see our website at: **www.aeronautbooks.com**.

I am looking for photographs of the less well-known German aircraft of WWI to complete this series. For questions or to help with photographs you may contact me at **jherris@me.com**.



ISBN: **978-1-953201-23-2**

Design and layout: Jack Herris
Cover design: Aaron Weaver
Digital photo editing: Jack Herris

Aeronaut Books
*Books for Enthusiasts by Enthusiasts*
www.aeronautbooks.com

# Table of Contents

The first Thomas-Morse MB-3 prototype, A.S.40092.

# Preface

Preceded by the MB-1 and MB-2 prototypes, the Thomas-Morse MB-3 single-seat pursuit was the first indigenous US airplane of this type. (The contemporary name for the single-seat fighter category of warplane was pursuit. Thus, pursuit will be used throughout this book.) First flying on February 21, 1919. Of all-wooden construction, the MB-3 was powered by a 300 h.p. Wright H-3 engine, which gave it a maximum speed of 225 km/h at sea level. Armament consisted of a pair of 7.62 mm machine guns. Even though harsh criticism was aimed at the MB-3 during trials at McCook Field, it was eventually selected as a standard single-seat pursuit for the United States Army Air Service (USAAS).

Apart from prototypes, a total of 61 MB-3s were built by Thomas-Morse, 50 for the Army Air Service and eleven for the USMC. A second contract for 200 improved MB-3As was awarded to Boeing, which had managed to considerably underbid Thomas-Morse. Securing this contract, the largest awarded in the immediate post-war years, secured the continued existence and growth of Boeing. It has subsequently been said that building the MB-3A provided Boeing with a practical lesson in how a pursuit should not be designed! Nevertheless, the MB-3/MB-3A holds the distinction of being the first indigenous US single-seat pursuits to enter quantity production. Including prototypes, a total of 265 were built.

The MB-3/MB-3A was the standard USAAS pursuit during the early and mid-1920s, serving with eight squadrons, replacing the mix of S.E.5as, SPAD XIIIs, and Fokker D.VIIs that had equipped the USAAS pursuit squadrons since 1919. The MB-3A was also successful in the air racing field.

Eleven MB-3s were also acquired by the United States Marine Corps (USMC). These airplanes did not remain in service for long, eventually being sold to the US Army Air Service. A number of MB-3As were subsequently modified as MB-3Ms, with the M suffix standing for 'modernized', serving as advanced trainers. Several MB-3A/MB-3Ms were used in the 1927 movie Wings, starring as French and German pursuits, with at least two being intentionally destroyed before the film cameras. Sadly, no MB-3/MB-3A was set aside for preservation.

Today, the Thomas Morse MB-3 series is virtually forgotten. To the author's knowledge, no publication solely devoted to the history of the Thomas Morse MB-3 series has appeared before. This book is an attempt to redress this imbalance. Apart from the MB-3/MB-3A, brief descriptions of the MB-1 and MB-2 pursuit prototypes have been included, as well as the MB-4, MB-6/R-2, MB-7 derivatives, as well as the TM-22/R-5 air racers. Also included are details on the Ordnance Engineering Corporation's Type D pursuit, the closest competitor of the MB-3, which also entered small-scale production and limited USAAS service.

## Abbreviations and Glossary

| | |
|---|---|
| AEF | American Expeditionary Force |
| ASTS | Air Service Tactical School |
| ASTTS | Air Service Technical Training School |
| C/n | Construction number |
| F.A.I.D. | Fairfield Intermediate Air Depot |
| GFE | Government-furnished Equipment |
| PG | Pursuit Group |
| PS | Pursuit Squadron |
| S/n | Serial Number |
| SS | School Squadron |
| USAAC | United States Army Air Corps |
| USAAS | United States Army Air Service |
| USMC | United States Marine Corps |

# Historical Background

When the USA joined forces with Great Britain and France against Germany on April 2, 1917, the US aeronautical industry was ill-prepared for war. A huge and very rapid expansion resulted in thousands of aircraft, Curtiss JN-4 primary trainers and British-designed DH 4 light bombers in particular, being churned out. Nevertheless, no less than 5,164 aircraft were acquired from British, French and Italian manufacturers for use by American frontline squadrons and for training purposes.

Out of this total, 1,566 were pursuits; including 38 RAF SE5as, 143 Sopwith Camels, five Sopwith Dolphins, 298 Nieuport 28s, 189 SPAD VIIs and no less than 893 SPAD XIIIs. An order for 1,000 SE5as was placed with Curtiss. In the end, adapting European aircraft designs to American production methods proved both difficult and time consuming, with Curtiss only managing to complete one SE5a in September 1918 before the production contract was cancelled. A plan to produce 3,000 SPAD pursuits in the USA was not realized.

Arriving over the Western Front in the late spring of 1918, the American Expeditionary Force (AEF) eventually fielded 20 pursuit squadrons. However, although US-built DH 4s reached the front, no pursuits of US manufacture were ready before the war ended on November 11, 1918.

# The Thomas-Morse Company

Formed in January 1917 through the merger of the Thomas Brothers Aeroplane Company (formed pre-war by two English brothers, William T Thomas and Oliver W Thomas) and the Morse Chain Works, the Thomas-Morse Aircraft Aircraft Corporation of Ithaca, New York, is today perhaps best known for their Thomas-Morse S-4 single-seat scout. First flown in November 1917, a total of 647 S-4s were built in two main variants, the S-4B and S-4C respectively. Powered by a rotary engine (either the 100 h.p. Gnome or 80 h.p. Le Rhône), the Thomas-Morse S-4 was deemed to have insufficient performance to match contemporary British, French and German pursuits. Nevertheless, the Thomas-Morse S-4 served as an advanced trainer, preparing pursuit pilots for the front. Post-war, many Thomas-Morse S-4s became movie stars, usually being painted in Hollywood-style fanciful markings representing different British (in particular the Sopwith Camel) as well as generic Great War German pursuits. Several Thomas-Morse S-4s have been preserved, with a few being in airworthy condition.

Even though the Thomas-Morse S-4 had been rejected as a frontline pursuit, the company developed two different types of pursuit prototypes in 1918, designated as the MB-1 and MB-2 respectively.

First flown in 1919, the MB-3 single-seat pursuit proved a success, with 65 being built by Thomas-Morse. Developments of the basic MB-3 included the two-seat MB-4 mail carrier, and the MB-6/R-2 and MB-7/R-5 air racers. However, a subsequent contract for 200 improved MB-3As was won by Boeing, resulting in Thomas-Morse being relegated to second-line status among US aircraft manufacturers. Despite producing several pursuit prototypes during the 1920s, including the parasol-wing MB-9, the all-metal TM-23 and the XP-13, none of these were accepted by the US Army Air Corps (USAAC). On August 5, 1929, Thomas-Morse was bought by Consolidated Aircraft. The Thomas-Morse Division was discontinued in 1934, thus ending nearly 20 years of aircraft development and production.

# The Thomas-Morse MB-1

**Above:** The Thomas-Morse MB-1 seen during ground trials. Note that no propeller seems to be fitted. Via Jack Herris

The MB-1 was a two-seat parasol-wing pursuit of mixed wood and metal construction, powered by a 400 h.p. geared Liberty 12 engine. The engine was transported to Ithaca in a guarded wagon. Work on the MB-1 was initiated in late 1917 by the British designer Benjamin Douglas Thomas. Although sharing the same last name, B D Thomas was not related to the Thomas brothers. Incidentally, B D Thomas' previous work included the Curtiss JN two-seat tractor biplane, better known as the 'Jenny'.

Two MB-1 prototypes were ordered, one of which was to be used for static tests. In order to achieve the lowest possible airframe structural weight, lightening holes were cut into the plywood bulkheads and various metal components. Amazingly, lightening holes were also included in the metal control column! This made the MB-1 airframe prone to structural failure. Nevertheless, the empty weight of the airframe amounted to 2,000 lb (908 kg), to which the 900 lb (408,6 kg) heavy engine was added. Presumably, this caused some problems with the centre of gravity. According to aviation historian R.L. Bliss, the MB-1 "resembled a dragonfly on steroids".

The armament consisted of four machine guns, including two synchronized forward-firing guns and two Lewis guns in a scarff mounting in the rear seat. Uniquely, a downwards-firing Winchester rifle was also carried. Even before the start of the flight trials, one of the two MB-1 prototypes suffered damage when the tail skid fitting broke off when the aircraft was parked in a hangar. The first flight attempt at the icy surface of Lake Cayuga near Ithaca in early 1918 ended in disaster when the undercarriage collapsed prior to take-off. Subsequently repaired, the prototype managed to struggle into the air but crashed moments after taking off. The two prototypes were then transferred to McCook Field, where they remained firmly grounded. Following arrival at McCook Field, the pair of MB-1s were not allotted serial numbers in the sequential numeric 'P' test series. MB-1 development was quickly abandoned in favour of the two-seat MB-2.

### Thomas-Morse MB-1 Technical Data

Wingspan: 37 ft (11,28 m)
Length: 22 ft (6,7 m)
Loaded weight: 2,375 lb (1,077 kg)

**Above:** The Thomas-Morse MB-1 prototype during engine trials. The abnormally large wing struts would surely have reduced the pilot's downward view. Via Jack Herris

**Above:** A three-quarter view of the Thomas-Morse MB-1. Via Jack Herris

**Above:** A front view of the Thomas-Morse MB-1 Via Jack Herris

**An MB-3A of the 17th Pursuit Squadron, the Snow Owls.**

# The Thomas-Morse MB-2

**Above:** A three-quarter front view of the MB-2. Via Jack Herris

The two-seat MB-2 was a biplane of orthodox design, having a fabric-covered wooden airframe with equal-span, unstaggered wings. Responsibility for design work was placed with B D Douglas. Power was provided by a 400 h.p. geared Liberty 12c engine, driving a huge four-blade propeller. The MB-2s intended armament was to consist of one 7.62 mm (0.3 in) and one 12.7 (0.5 in) synchronized forward-firing machine guns and one 7.62 mm (0.3 in) machine gun in a swivelling mount in the rear cockpit.

As with the MB-1, efforts were made to reduce overall weight of the MB-2. Two MB-2 prototypes were ordered, USAAS serial numbers A.S.25805 and A.S.25806, with the latter taking to the air in November 1918. (Some sources, including The American Spad by R.L. Bliss, state that the MB-2 was never flown.) Apparently, the second MB-2 did not proceed beyond an incomplete fuselage.
The first MB-2 prototype featured single-bay wings. However, static tests showed the wings being of insufficient strength. Although twin-bay wings were fitted, as well as a 450 h.p. Liberty engine along with a different radiator and a four-blade propeller, MB-2 development was abandoned. No performance data for the MB-2 appear to have survived.

### Thomas-Morse MB-2 Technical Data

Wingspan: 31 ft (9,45 m)
Length: 24 ft (7,31 m)
Height: 8 ft (2,43 m)
Wing area: 323 sq ft (30 m$^2$)
Empty weight: 2,047 lb (929 kg)
Loaded weight: 2,773 lb (1,258 kg)
Fuel capacity: 50 USgal (1,891 l)

**Above:** The MB-2 prototype seen just after completion. Via Jack Herris

**Above:** One of the MB-2 prototypes seen during static tests. Unfortunately, the date on the sign is illegible. Via Jack Herris

**Above:** An excellent view of MB-2 A.S.25806. Via Jack Herris

**Above:** A three-quarter side view of MB-2 A.S.25806. Via Colin Owers

# The Thomas-Morse MB-3

**Above:** The second MB-3 prototype, A.S.40093, after arrival at McCook Field. Via Jack Herris

Clearly inspired by contemporary European pursuits, most particularly the French SPAD XIII, the MB-3 in fact resembled a scaled-down single-seat variant of the earlier MB-2. Design responsibility yet again rested on B D Thomas. The MB-3 promised to be a high performance pursuit, with maximum speed of 150 mph (241 km/h) and an initial climb rate of 1,500 ft (457 m)/min being envisaged. Using the R.A.F. 32 airfoil, Thomas created a single-seat pursuit of orthodox fabric-covered wooden construction. Power was provided by a 300 h.p. liquid-cooled Wright H (a licence-built Hispano-Suiza) engine, which, despite having a tendency to vibrate badly, was deemed as being suitable for pursuit aircraft. A contract for four MB-3 prototypes, A.S.40092-40095, was signed on September 27, 1918, at a cost of $ 66,000. However, Thomas-Morse was not the sole manufacturer struggling to secure contracts for the new Air Service single-seat pursuit. In fact, prototypes of no less than six different pursuits were entered into the fray.

Apart from the MB-3 these included the Pomilio FVL-8, the Lewis & Vought VE-8, the Engineering Division Verville VCP-1, the Loening M-8 and the Ordnance Engineering Corporation Type D. Out of these, only the MB-3 and the Ordnance Engineering Corporation Type D secured production contracts.

The first MB-3 prototype was first flown on February 21, 1919 by Frank H. Burnside. Overall, the MB-3 performance was deemed as excellent, with the aircraft being able to climb "almost vertically" to an altitude of 2,000 ft (610 metres). 10,000 ft (3,048 metres) was reached within less than five minutes. The four prototypes carried an armament of twin 0·3 in (7,62 mm) machine guns, mounted in the upper front fuselage. The first prototype, A.S.40092, subsequently underwent destructive static tests at McCook Field in April 1919. These tests indicated that the MB-3 airframe did indeed possess sufficient strength.

The second prototype, A.S.40093 (McCook Field serial number P66), was despatched to

**Above:** One of the MB-3 prototypes at Ithaca. The abundance of wheel chocks may indicate an engine test. Via Colin Owers

McCook Field for trials. The 'McCook Field MB-3 Evaluation report' was submitted on February 20, 1920. Although possessing excellent speed and climb performance, the MB-3 was also criticized on a number points:

"The airplane is easily taxied; is steady and easily controlled on take-off, and leaves the ground quickly.

The airplane is well balanced and controls are normal in action.

The airplane maneuvers well in all positions.

The pilot has poor vision in the air, his head being too far below the level of the top wing. View is all right for landing.

In landing the airplane is easy to hold off and has no tendency to ground spin or nose over.

Cockpit is entirely too small. It is badly cramped in every way, even for pilots of small stature. This makes it difficult to locate motor controls conveniently.

This airplane has caused an endless amount of work from the maintenance standpoint, a large share of which was due to the fuel installation. After nearly every flight the main tank developed leaks, due to its complicated shape comprising many sharp corners. Also the tank had a very flimsy support, resulting in its breaking through the bottom of the fuselage after about 10 hours' flying. This was due also in part to the trapping of the overflow oil from oil tank between gas tank and bottom of fuselage, causing the veneer to rot out. A great deal of trouble was experienced with the cooling system until a new radiator was installed with different method of piping. There was no fire wall between motor and gas tanks.

On the new Thomas-Morse MB-3 (the fourth prototype), an attempt has been made to remedy the majority of the faults mentioned above. The cockpit has been lengthened but is still entirely too narrow. A new fuel installation has been made which has already given endless trouble through its gas pumps, check valves, bypasses, etc., apparently none of which had been tested or tried.

After less than two hours' flying one of the main fittings in the wing section of the fuselage was found broken. In replacing the fitting it was discovered that it was necessary to make up special wrenches in order to tighten up certain brace rods in this section of the fuselage.

It was found necessary to cut holes in the engine

**Above:** The second MB-3 prototype, A.S.4093, at McCook Field. Note the unusually long pitot tube. Via Jack Herris

section of fuselage in order to get at magnetos.

The new airplane, being slightly heavier, does not maneuver quite as rapidly as the old one. The new shutter control when closed throws the airplane out of balance, making it slightly tail heavy.

This new airplane has not been flown enough to get definitive information as to maintenance.

(Signed) J.M. Johnson,
Test Pilot.

In the event, following necessary modifications, Thomas-Morse secured a contract (No. 265) on June 19, 1920 for 50 MB-3s plus spares, a mere three days after an order for a similar quantity of Orenco Type Ds. The cost for the MB-3 contract amounted to $720,432.75.

The production MB-3s differed from the prototypes in a number of respects. In response to the McCook Field report, the fuselage had been lengthened somewhat (albeit not widened), the wing-mounted radiator also being slightly off-set to starboard from the fuselage centreline, making it possible to fit a 12gallon gravity fuel tank. The vertical fin was also cut down in size. These modifications resulted in an empty weight increase of 130 lb.

By early 1921, MB-3 production was progressing rapidly. Tentative plans for the distribution of the first 14 production aircraft included four to be transferred to McCook Field, two for the Air Service Technical School, based at Chanute Field, and eight for the Field Officers School, based at Langley Field. The remaining 36 MB-3s were to be delivered to the 1st Pursuit Group at Kelly Field.

However, the MB-3 troubles were far from over. Having discovered handling as well as engine cooling problems in the first production MB-3, Lieutenant J.A. Macready took off from Ithaca in the eleventh production MB-3, A.S.63341, on March 30, 1921. During a high speed diving test with the engine merely ticking over, the leading edge of the upper wing caved in. Lieutenant Macready was lucky to get down in one piece, with his MB-3 overturning during the forced landing.

Macready's report on the accident stated: "One of my duties was to test the maneuverability as a fighting airplane. The dive on the enemy is one of the primary functions of combat. With this in mind, and also to note the effectiveness of the closed

**Above:** A rare air-to-air view of MB-3A A.S.68418. This was one of the last 50 production aircraft, as evidenced by the revised fin. The purpose of the numeral 9 is unclear. Via Colin Owers

shutters, I dove the plane at a very steep angle with the power off. I was carefully studying the airplane as it dove, intending to keep a margin of safety and refraining from any quick movements of the controls that would cause undue strains or stresses.

Suddenly I heard a terrific report or crashing sound and felt the plane heavily jarred. I thought that the propeller had broken and crashed into a wing. The pilot can see the top of the top wing from his seat. Looking up, I observed that practically the entire top of the wing was gone, and that webs and woodwork was bare with the fabric stripped from them. The entire leading edge of the wing had collapsed and stripped the top fabric from the wing, leaving it with almost no lift or sustaining power."

Miraculously, Macready only suffered a few cuts and bruises. After extracting himself from the wreckage, Macready ordered guards placed close to the MB-3 until it could be thoroughly inspected by an Engineering Division representative. He also called McCook Field, relating details of his accident. Incidentally, the problem of the upper wing fabric covering being ripped off in a dive was not an unknown event. During 1918, several such cases occurred in Nieuport 28s flown by US pilots over the Western Front.

Macready's near-fatal crash resulted in prolonged trials involving strengthening the upper wing. Series production by Thomas-Morse was halted until these problems had been overcome, not re-commencing until early 1922. Nevertheless, on May 11, 1921, Thomas-Morse secured a supplemental $ 26,396.21 contract for strengthening the wings of "last 49 airplanes and spares." Another contract, dated October 21, 1921, called for "reinforce wings of last 36 airplanes and 9 sets of spare upper wings (and) make 16 other changes." This second contract amounted to $ 88,100.00. It seemed as if Thomas-Morse would indeed become the Air Service's primary supplier of pursuit airplanes.

The task of investigating Macready's accident was assigned to the National Advisory Committee for Aeronautics (NACA). Established in 1915, NACA was 'to supervise and direct the scientific

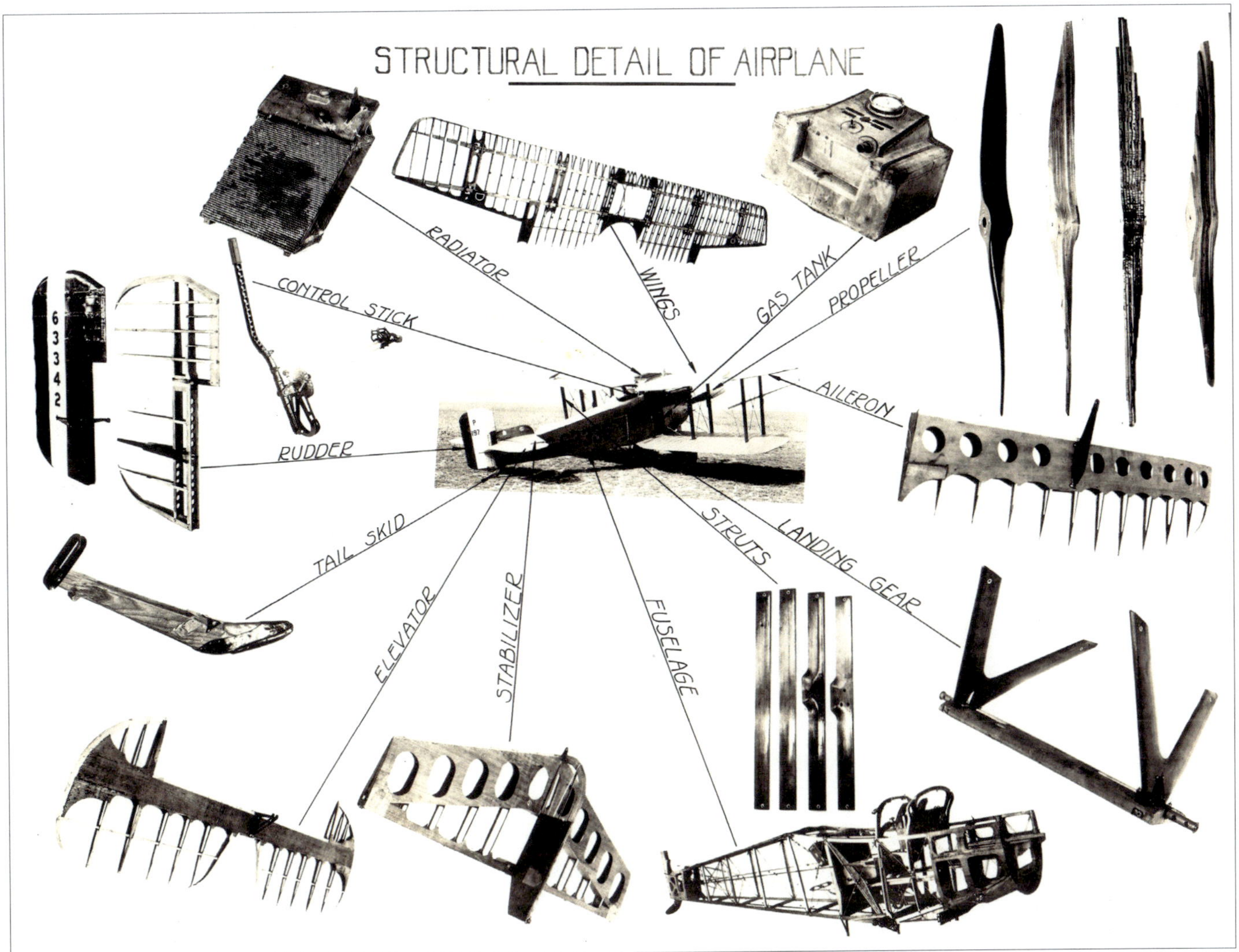


**Above:** An excellent overview of the various MB-3 sub-assemblies. The MB-3 is A.S.63336, marked as P197 at McCook Field. However, the rudder on the left is marked A.S.63342! Also of interest is the various stages of propeller construction in the upper right-hand corner. Via Colin Owers

**Above:** Although carrying Air Service insignia on the upper surfaces of the wings and rudder, the serial number of this MB-3 has yet to be applied. Via Colin Owers

**Above:** An MB-3A, presumably seen with Boeing prior to delivery. Of note is the Fokker D.VII in the background. Via Colin Owers

study of the problems of flight with a view to their practical solution", as well as to "direct and conduct research and experiment in aeronautics'. In 1919, NACA moved to Langley Field (named after Samuel Pierpoint Langley of Aerodrome fame) north of Hampton, Virginia.

One MB-3, A.S.63335, the fifth production airplane, was loaned to NACA in January 1923. While at Langley Field, the serial number NACA 8 was assigned to the MB-3. The subsequent investigation carried out resulted in 'Report No. 193. Pressure Distribution Over The Wings Of An MB-3 Airplane In Flight'. According to the report, in order to establish pressure distribution over the wings during different flight modes, holes in the surface of the wings were connected to multiple recording manometers located in the fuselage. The results were recorded photographically through 30 diaphragm capsules. The report states, in part: "The wing section is shown in Figure I together with the R.A.F. 15 section for comparison. It is very interesting to note the great divergence between the actual section turned out by the contractor and the R.A.F. 15 section which was supposed to be used. The change was probably made after the original design was laid out to accommodate deeper spars, but instead of adopting a thick, but still efficient section, the upper surface of the R.A.F. 15 was simply bulged out over the spars. The resulting section undoubtedly gives a high-speed performance distinctly inferior to the R.A.F. 15.

"It was considered desirable to make a number of changes in the standard airplane, first from the point of safety, and second to facilitate the test. The more important are enumerated below:

1. The radiator and the fuel tank were removed from the center section, which was made to conform with the wing section. This was done in order to prevent disturbance of the air flow in this section of the upper wing, to provide greater visibility for the pilot, and to permit loading the manometers with film conveniently.
2. A 180 H.P. Lamplin radiator was placed just over the axle and was found to give very satisfactory cooling.
3. The rear center section bulkhead was changed so that it aligned with the rear center section strut, both to allow more room for placing the multiple manometers and to give greater strength and rigidity to the center section.
4. A number of heavy ribs were put in both the upper and lower wings, as several wing

**Above & Below:** The Lieutenant John J Mitchell Trophy Race at Selfridge Field on October 14, 1922 saw the participation of six MB-3s. Seen here is Lieutenant Donald F Stace's Boeing MB-3, individual Race Number 54, taking off. Note the different markings on each side of the fuselage. Via Colin Owers

**Above:** Capt. O.W. Broberg taking off during the Lieutenant John J Mitchell Trophy Race on October 14, 1922. Given his surname, Broberg possibly had Swedish ancestry. Broberg ended up in third place. Via Colin Owers

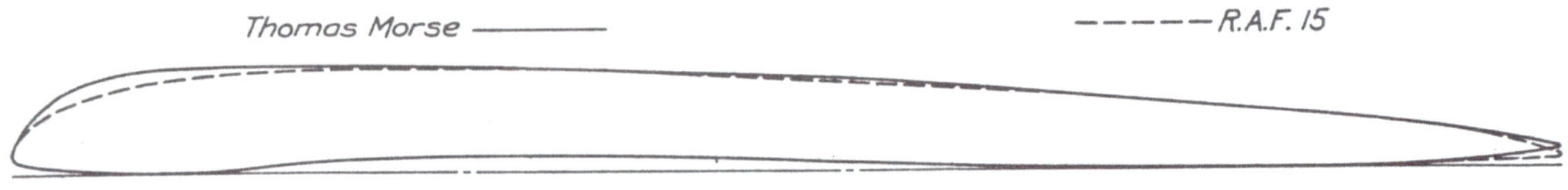


Fig. 1.—Comparison of the Thomas Morse section with the R. A. F. 15

**Above:** A contemporary, comparative drawing showing the difference between the RAF15 and MB-3 airfoil. Originally published in F.H. Norton's report Pressure Distribution Over the Wings of an MB-3 Airplane in Flight. *Skyways* No. 22 (April 1992)

failures on this type of airplane indicated insufficient strength here.

5. When the wings were re-covered, the stitching was closely placed to prevent the fabric's stripping.
6. Heavier interplane struts were installed to prevent lateral deflection.
7. A number of fittings were replaced by ones of heavier metal and the engine section was stiffened.
8. The tip of the balance on the elevator was removed to prevent hunting of the longitudinal controls.
9. The rudder post was stiffened to prevent vibration.
10. All of the military equipment was removed to make room for the instruments.
11. A four-bladed propeller, which was put on the airplane, somewhat reduced the vibration.

"Such extensive changes, of course, took a considerable length of time, but it was felt that they were justified because the nature of the present test demanded very violent maneuvering and the instruments installed required a minimum of vibration. The pilot reported that the airplane as rebuilt could be handled easily and was a decided improvement over the original model."

The report's conclusions recommended an overall increase in the strength of the wings to avoid the fabric being ripped off. The lower surfaces of the leading and trailing edges "should be stiffened."

On November 1, 1923, the MB-3 was returned to the USAAS.

In the end, the high-performance fabric-covered biplane would soon be replaced with all-metal designs, thus making the report's recommendations obsolete. Even though the majority of the wing area of such airplanes would remain fabric-covered, the leading edge was usually shaped out of duralumin, which meant that the wing had a greater overall strength.

### Thomas-Morse MB-3 Technical Data

Source: *Jane's All the World's Aircraft*, p. 482a
Type of machine: Single-seater biplane.
Name or type No. Of machine: M.B.3.
Purpose for which intended: Fighting.
Span: 26 ft.
Overall length: 19 ft.
Maximum height: 8 ft
Chord: 5 ft. 3 in.
Total surface of wings: 250 sq. Ft.
Engine type and h.p.: 300 h.p. Hispano-Suiza.
Weight of machine empty: 1,360 lbs.
Load per sq. ft.: 8 lbs.
Petrol tank capacity in gallons: 65 gallons
Performance:
Speed low down: 164 m.p.h.
Landing speed: 55 m.p.h.
Climb:
To 1,000 feet: 20 secs.
To 10,000 feet in minutes: 4 mins. 52 secs.
Total weight of machine loaded: 2,000 lbs.
Two Browning machine guns
Useful load (with water): 800 lbs.

## The Ordnance Engineering Corporation Type D

Before continuing with the MB-3 story, it may be of some interest to briefly describe its closest competitor for the role as the Air Service's first indigenous pursuit; the Ordnance Engineering Corporation Type D.

Formed at Baldwin, Long Island, in the summer of 1916, the Ordnance Engineering Corporation's first offering was a side-by-side two-seat biplane trainer. Known as the Type A, the trainer remained a prototype. In June, or possibly July, the Ordnance Engineering Corporation hired the services of Etienne Dormoy, an experienced aircraft engineer who had previously worked for Société Pour Aviation et ses Dérivés (SPAD). By the summer of 1917, Dormoy was a member of the French Aeronautical Mission to the USA. Dormoy's first task was to design a pursuit. Appearing in early 1918, the Type

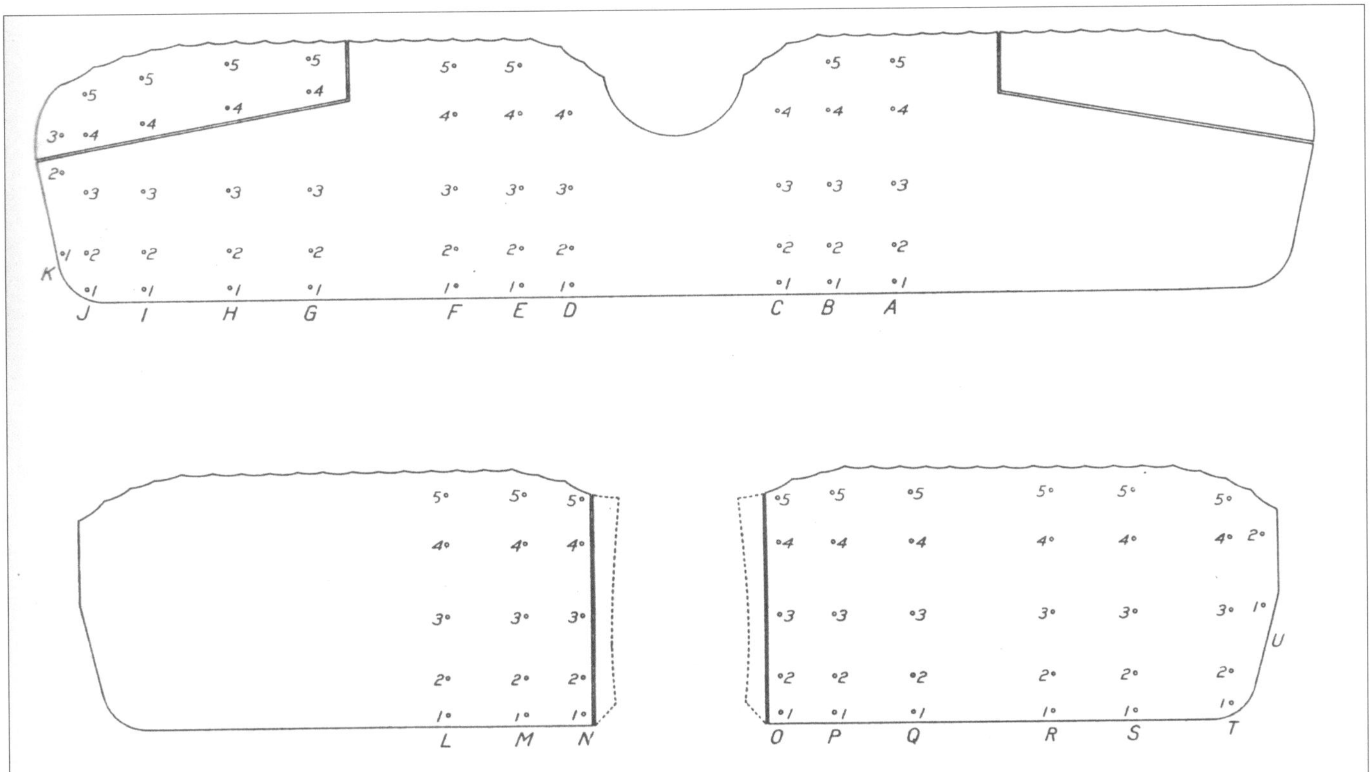


Fig. 2. — Plan of wings showing location of pressure holes.

**Above:** A contemporary drawing showing the location of the pressure holes. Originally published in F.H. Norton's report Pressure Distribution Over the Wings of an MB-3 Airplane in Flight. *Skyways* No. 22 (April 1992)
**Below:** A contemporary drawing showing the distribution of air pressure in flight. Originally published in F.H. Norton's report Pressure Distribution Over the Wings of an MB-3 Airplane in Flight. *Skyways* No. 22 (April 1992)

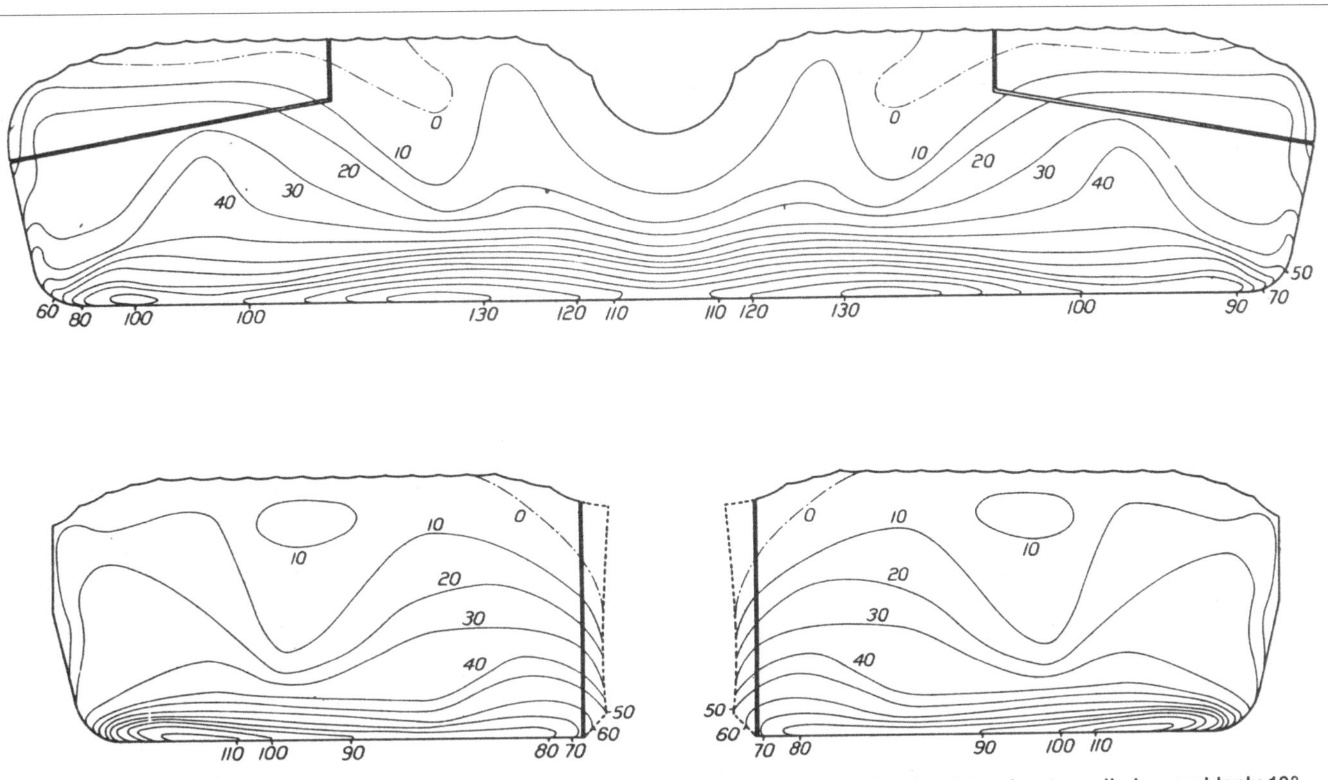


Fig. 3. — Lift of wings in a sudden flattening out of a dive at 140 mph & 1900 RPM. Acceleration 3.6g; elevator pulled up suddenly 10°.

**Above:** A prototype Ordnance Engineering model D fighter undergoing flight testing on April 4, 1919. A long pitot tube is on the right outboard wing struts.
**Below:** Ordnance Engineering model D serial 40108 (P-67) at McCook Field, Dayton, Ohio, in 1919. This second aircraft was the first delivered with its 300 hp Wright-Martin Hispano engine.

B biplane pursuit was powered by a 160 h.p. Gnome Monosoupape rotary engine. The armament was to consist of three 7.62 (0.3 in) Marlin machine guns. Four Type Bs were ordered, along with five Type C trainer derivatives. However, the USAAS decided against ordering any further Type B pursuits. Nevertheless, the Orenco Engineering Corporation continued work on the Type D pursuit. Powered by a 300 h.p. Hispano-Suiza H 8-cylinder water-cooled engine, the Type D was of wooden construction. The fuselage was plywood-covered, with the two-bay equal-span wings being fabric-covered. Incidentally, in early 1919 the Ordnance Engineering Corporation coined the acronym Orenco. This was made in order to avoid confusion with the Army Ordnance Branch.

Four Orenco Type Ds (serial numbers A.S.40107–40110) were ordered for evaluation at McCook Field. The Type D received high marks, with test pilot

**Above:** A somewhat fuzzy ground-to-air view of Capt. Guidera's MB-3 during the Lieutenant John J Mitchell Trophy Race. Guidera ended up in second place. Via Colin Owers

Clarence B. Coombs stating:

Airplane has good flying qualities, responds very well both taking off and landing. Is slightly tail heavy in climbing positions up to 20,000 feet. Landing qualities are better than normal with single-seaters of this type, the airplane rolling about 600 feet on normal landing. It does not tend to nose over, though it tends to porpoise badly if landed on rough ground, or at high speed. It does not taxy easily, the rudder being small, and the weight on the tail skid excessive. Pilot's vision is better than in the average machine of this type, the top wing offering practically no impediment to vision. View directly ahead is partly impeded due to the depth of the fuselage. Locations of the controls are exceptionally good. The ailerons on this airplane are insufficient in surface, though this is being taken care of in the new wing design. The shutters are very efficient, the radiator being of ample size to cool under any conditions. Some considerable trouble has been experienced with the oil tank, the rivets pulling out. Maintenance has been difficult due to the inaccessibility of certain vital engine parts, this is being taken care of in the new model. Alignment is easy to maintain, no trouble being experienced except the loosening of inner bays. Practically no vibration is noticeable at cruising or wide throttle positions. It has an exceedingly wide speed range, and responds rapidly in combat maneuvers (this plane has not been spun). This airplane handles better than the Sopwith Camel and Snipe, the Thomas-Morse, Nieuport and Morane Parasol rotaries, the Spad and S.V.A."

At the time, when prototypes of a new military airplane were acquired by the War Department, this also included design rights. As a result, the company which had designed the airplane in question, could not feel secure in obtaining production orders. Instead, bids were requested from different manufacturers. For some manufacturers, including Orenco and Thomas-Morse, this system was to have dire consequences.

In the event, Curtiss underbid Orenco, receiving a contract (No. 262) for 50 pursuits on June 16, 1920. The Air Service serial numbers were A.S.63281–63330. The cost amounted to $524,000 plus an additional $55,167.37 for spares. The third Type D prototype, McCook Field serial number P-101, was shipped to Curtiss for use as a pattern aircraft. Curtiss eventually delivered a slightly redesigned pursuit, including wing dihedral, balanced elephant ear ailerons, and revised engine installation. Despite losing an important production order, Orenco persisted in developing the basic Type D. Known as the Type D2, this had unequal-span, single-bay wings. Three were ordered by the Air Service as the PW-3 (Pursuit, Water-cooled engine Type 3). However, the PW-3s remained grounded, being subjected to static stress and fatigue tests. It was soon discovered that the aircraft were structurally

**Above:** Boeing MB-3A A.S.68586, McCook Field P-274, was experimentally fitted with faired lights into wing tips for night landings. The airplane is marked as BOEING MB-3A on the rudder. Via Colin Owers

deficient, as well as displaying crude workmanship.

Naturally, the Curtiss-built pursuit Orenco D also underwent tests at McCook Field. One major criticism was engine vibration, with the veneer-covered fuselage also vibrating badly. Nevertheless, it was decided that 54 Orenco Ds were to be delivered to Kelly Field outside San Antonio, Texas, for the 1st Pursuit Group. On July 19, 1921, 1st Lt Willard S. Clark was killed in the sixth production Orenco D. The cause of the crash was due to that "poor workmanship was found in the brazed wing strut fittings, and the strut bolts were of very poor material (lowest grade Bessemer screw stock) and the use of such material in places subject to stress 'is a criminal procedure'", with poor workmanship also being severely criticized. Other accidents resulted in a thorough inspection of the Orenco D fleet. The inspection report listed more than 30 points, which, along with a discussion with the Commander of the 1st Pursuit Group, Major Carl E. Spaatz concerning the pilots lack of trust in the aircraft, resulted in a directive, dated May 22, 1922. This stated that out of the 20 Orenco Ds in storage, four were to be transferred to ROTC units for instructional purposes, and the remaining 16 destroyed.

That Orenco failed to secure the production contract for its Type D pursuit, virtually spelled the end of the company's involvement in aviation. Although further designs were being worked on in the immediate post-war years, Orenco dropped out of the aircraft manufacturing business in 1922. (Incidentally, one Orenco F tourer went on to have a long Hollywood career. Remarkably, this airplane still survives.)

### The Ordnance Engineering Corporation Type D Technical Data and Performance Characteristics

Wingspan: 30 ft (9,14 m)
Length: 21 ft 6 in (6,55 m)
Height: 8 ft 3 in (2,52 m)
Wing area: 261 sq ft (24,24m$^2$)
Empty weight: 1,666 lb (756 kg)
Gross weight: 2,432 lb (1,103 kg)
Maximum speed at 5,000 ft (1,524 m): 144 mph (232 km/h)
Service ceiling: 22,000 ft (6,706 m)
Range: 275 miles (442 km)
Fuel capacity: 55USgal (208 l)
Armament: 2 x .30 in machine guns

## Pilot Opinions

Although it would seem that no proper comparative evaluation between the Orenco D and the MB-3 actually took place, the advantages (and disadvantages) of the respective pursuits are clearly revealed in a letter dated September 1921. The letter, written by test pilot Lt John A. Macready, states: "The Thomas-Morse MB-3 and Ordnance D have been officially decided upon as the standard pursuit types for the United States Army primarily because of the excellent results obtained from their performance tests. There are other inherent qualities of these airplanes which make them of less value than seemingly inferior ones. An airplane may

**Above:** Boeing MB-3A A.S.68586, McCook Field P-274, equipped with faired lights into wing tips for night landings. McCook photo 18516 via Colin Owers

**Above:** Front view of Boeing MB-3A A.S.68586, McCook Field P-274, equipped with faired lights into the wing tips for night landings. The results of these trials are unknown. McCook photo18515 via Colin Owers

have a wonderful performance and yet may not be able to use these qualities to the best advantage in maneuvering. There is a tendency for them to lose their speed in turns, especially in climbing turns which are among the most important airplane performance qualities of combat. This quality seems to be more or less characteristic of airplanes which are inherently unstable and which have a very quick and responsive snap to their controls. The Thomas-Morse MB-3, Ordnance D-1, and Packard-Fokker (P195) have excellent performance in straight-away flight, but the bottom appears to fall out of them in maneuvering. An opposite type to these is the Loening Monoplane. The value of this plane is yet unproven… In quickness and snap, in turning and maneuvering, the Thomas-Morse, Ordnance, and Packard-Fokker are far better, but so much speed and altitude are lost in making these maneuvers that their efficiency is greatly impaired. In a maneuver with the Thomas-Morse the air speed will drop from 140 mph to 60 or 70 mph within a period of one or two seconds, while the Loening will only drop from approximately 126 mph to 90 mph… Maintenance issues features are also very

**Above:** Boeing MB-3A A.S.68586, McCook Field P-274, equipped with faired lights into wing tips for night landings. McCook photo 18517 via Colin Owers

important. The simpler an airplane is the better it should be from the maintenance standpoint. The Thomas-Morse and Ordnance are extremly (sic) unsatisfactory airplanes in this respect.

"One very valuable quality of pursuit airplanes is their ability to remain steady in flight. The Thomas-Morse and Ordnance are unstable airplanes, which bob around, and constantly change their position in a dive or when maneuvering. This is a distinctly unsatisfactory condition from the standpoint of a gunner about to fire on an enemy airplane. The airplane must be steady in order that the pilot's guns may be turned on the enemy with accuracy. An airplane that is constantly changing its position when its nose is pointed at the enemy airplane is not a good combat airplane, for the reason that accuracy of fire is very difficult… Airplanes such as the Thomas-Morse and Ordnance will never be liked by average pilots in the field because of their undesirable flying qualities, and their instability. These airplanes will be flown by pilots originally, but after the first time difficulty will be encountered in getting them to fly them. They will not fly them for pleasure. If ordered to they will fly them as a duty, but they will have no compunction when the airplanes are damaged, or in pointing out all the undesirable features that they possibly can.

"Mechanics will work in somewhat the same way. The inaccessible and difficult maintenance will make them dislike airplanes of these types. An airplane such as the Loening, with easy maintenance, and comparatively easy flying qualities, would be liked by both pilots and mechanics." The monoplane Loening PW-2A (a development of the the M-8) preferred by Macready was lost on October 20, 1922, due to wing flutter. The post-crash investigation stated: "that the aerodynamic phenomena in a wing which is weak in torsion are not clearly understood, and in this particular instance has led to disastrous results."

## An Improved Pursuit, the MB-3A

On February 21, 1921, bids from six manufacturers were received for producing either batches of 50 or the entire contract of 200 aircraft. The highest bidder was Dayton-Wright ($2,201,000), followed by L.W.F. ($2,133,000), Curtiss ($1,982,000), Thomas-Morse ($1,926,000), Aeromarine ($1,832,000), and Boeing ($1,488,000). With War Department funding being substantially lower than during wartime, spending as little as possible on as much equipment (in this case, pursuit aircraft) as possible opened the door for Boeing.

For the War Department, it was a question of dividing the order among several manufacturers, thus ensuring the short-term survival at least of some companies in a highly-competitive field. This would be more expensive than awarding the entire order to a single manufacturer. Being on the brink of survival, the small Seattle-based Boeing would make the utmost to secure the order:

"The cold word from Dayton in January (1921)

**Above:** An air-to-air photo of Thomas-Morse MB-3A A.S.68332, individual No.52, with hornet insignia of the 43rd School Squadron, Kelly Field, Texas. The photo was taken on April 20, 1926. Note FAID on rudder stripes indicating that the airframe was overhauled at the Fairfield Air Intermediate Depot, Fairfield, Ohio. Via Colin Owers

was that there wasn't enough money to keep even a small part of the airplane industry alive. It was going to be a case of survival of the fittest. The L.W.F. Company was in the hands of receivers already… The only new planes the Army could buy were two-hundred MB-3 type pursuits designed by the Thomas Morse Company, Egtvedt learned on a Dayton visit.

"Everybody is going to be scrapping for that order," said Major Fleet (the McCook Field contracting officer).

"Will you give all two hundred to one company?" asked Egtvedt.

"We're getting bids on quantities of fifty to two hundred."

In Seattle the challenge was taken up. They'd have to bid low, or there was no use bidding. Phil Johnson unrolled the drawings and called in the shop foremen, one by one, to get their ideas as to the amount of work on the various parts. Then they met in Gott's office. When the estimates were added, the price came to $ 10,175 per plane for fifty; $ 6,617 for two hundred.

Ed Gott (Boeing V.P.) went to Dayton for the bid opening. Major Fleet began reading the bids: "Aero Marine, $1,832,000; Boeing, $1,448,000; Curtiss, $1,982,000; Dayton-Wright, $2,201,000." Gott was concerned. "We must be off," he thought.

"L.W.F., $2.133,000; Thomas Morse, $1,936,000." Gott was now thoroughly alarmed. Thomas Morse had built the airplane. Morse's figure was $478,000 higher than Boeing's. A $400,000 loss would mean disaster.

"Boeing is low at $1,448,000."

A silence followed. Gott looked around. Some of the men were glaring at him. When the meeting broke up, none of them came over to congratulate him. They gathered in small groups. "He'll lose his shirt," someone was saying. A Curtiss man came up with a friendly air. "You can withdraw," he counselled. Gott went to the Miami Hotel and called Phil Johnson. "Everyone says we'll go broke at these figures. Think we should withdraw?"

"We're in business," boomed Phil. "Let's stick to our guns. We can build 'em for that, easy."

Boeing backed Phil up. But there was still a question about the bid being accepted. The whole thing was now in the hands of Washington. The pressure was on General Menoher, chief of the Air Service, to split the business. There were reports that it was up to Secretary of War Weeks, that it was up to President Harding, that the whole project might be dropped for another year while the Secretary of War and Secretary of Navy resurveyed the arms program.

Five weeks had elapsed when Ed Gott, beetled

**Above, Below, & Facing Page:** Three walk-around photos of MB-3 No.63334 at Ithaca, New York. This was the fourth production MB-3. Via Colin Owers

and drawn, went to Secretary Weeks.

"I will have made my decision by next Saturday," the Secretary said. "The question is whether it is worth an extra $800,000 for us to split the business up. I don't think it is. I don't think the President will favor it. But I must look into it thoroughly."

On April 21, 1921, the $1,837,626.38 contract was awarded to Boeing. This included a total of $164,000 to cover the cost of spares and support gear, as well as crating, etc. As noted in the bid, the cost for each airframe was $6,617, less government-furnished equipment (GFE) such as engine, armament, and instruments. No Boeing model number was assigned to the MB-3A.

At the same time, the investigation into the cause behind the upper wing failure suffered by Lieutenant Macready was under way by NACA. Until the exact cause and remedy had been established, production by Thomas-Morse of the MB-3 would not begin in earnest until early 1922, with first delivery taking place on July 29. The month before, Boeing had completed its first, improved MB-3A. The last MB-3A, A.S.68436, was delivered on December 27, 1922. For the US aircraft manufacturers, securing a single contract for 200 military aircraft would remain unsurpassed for the next seventeen years. The decision to award the MB-3A contract to a single company did not sit well with the other bidders.

In late 1921, Boeing began work on the MB-3A. Series production was still some way off, though. Compared to the earlier MB-3, the MB-3A featured a number of structural changes. These included relocating the radiator from the upper centre wing to the fuselage sides. Fuel capacity was also increased. The improved Wright H-3 engine developed 320 h.p. A redesigned empennage was fitted to the last 50 production aircraft (this was not retro-fitted to earlier aircraft). According to a Change Order, dated September 11, 1922, the revised empennages would be fitted to the last 100 production aircraft. The continuous engine cooling problems led to McCook Field using the second production MB-3, A.S.63332 (P164) to change the location of the radiator from the upper centre wing to two fuselage-mounted radiators.

One sample MB-3 was sent to Boeing, but this was soon written off at Seattle in a landing accident. With Boeing's own airfield being insufficient, it was decided to perform test flights of completed aircraft using the parade ground at Camp Lewis, an army base located some 30 kilometres from Seattle. The first flight of a Boeing-built MB-3A took place on June 7, 1922. Unfortunately, the pilot hit a small ditch upon landing, resulting in the

**Above:** The wreck of MB-3 No.63341 following Lieutenant Macready's accident on March 30, 1921. Via Colin Owers

**Below:** A close-up of the rear fuselage structure on MB-3M A.S.63342. Via Colin Owers

**Above:** Many construction details, such as shape and position of the inspection panel and the 'LIFT HERE' texts, are discernible in this photo of MB-3M A.S.63342. Via Colin Owers

**Below:** An interesting close-up of the cockpit area of MB-3M A.S.63342. Of note is the small windscreen, and what appears to be a directional gyro in front of it. Via Colin Owers

**Above:** General W 'Billy' Mitchell with his Boeing MB-3A A.S.68264. Named *Hawk*, Mitchell's MB-3A was finished overall jet black, also being adorned with his personal insignia. Via Colin Owers

aircraft overturning in front of the assembled Boeing and Army dignitaries. However, this inauspicious beginning did not matter much.

According to the Engineering Division at McCook Field, the MB-3A "has provided the Air Service with a new improved Type I pursuit airplane from which the faults of the earlier design, the Thomas-Morse MB-3, have been carefully eliminated."

Problems with the MB-3A persisted though, with one factory-fresh airplane being lost on November 5, 1922. During a test flight, Frank Tyndall had to bail out due to structural failure. In the process, Tyndall became the second US member of the Caterpillar Club.

General Billy Mitchell, who flew his personal MB-3A, A.S.68264, polished jet black and nick-named *Hawk*, the MB-3A was "As good as if not superior to any pursuit type in the world."

Incidentally, development of an improved variant, unofficially known as the MB-3B, nearly reached fruition. The MB-3B would have featured a fuselage lengthened by two feet to improve directional and longitudinal stability, as well as ailerons on all four wings. In retrospect, it may be considered a good thing that the MB-3B never materialized. With the all-wood airframe of the MB-3A causing both production and maintenance problems, continuing the development of an already obsolete design would have been a bad idea. Boeing strongly believed that an improved pursuit would be required by the Air Service. Initially, the use of duralumin, as used by Thomas-Morse in their TM-23 pursuit, was considered but rejected. The result was the Boeing 15, which, construction-wise, was greatly influenced by the Fokker D.VII. Interestingly, three Fokker D.VIIs had been transferred to Boeing in 1921. It may be recalled that the Fokker D.VII was considered the best pursuit of the Great War. The fuselage of the Fokker D.VII consisted of a welded steel-tube frame, which greatly contributed to the aircraft's inherent strength and ease of construction. Both the fuselage and the wooden wings were fabric-covered. One of the Fokker D.VIIs was flight tested, with the other two being subjected to destructive static testing. It can be said with ample justification that the Fokker D.VII provided the source of inspiration for the long line of pursuits developed by Boeing during the mid-to-late 1920s. The Boeing 15 was to be powered by a 435 h.p. Curtiss D-12 engine, having a welded steel-tube fuselage and tapered wings. The preliminary design of the Boeing 15 was finalized on January 10, 1923.

It is worth noting that on September 27, 1923, Boeing was awarded a contract for three PW-9 pursuits for testing at McCook Field. The Air Service subsequently placed several orders for developed variants of the basic PW-9. Between the mid-1920s and early 1930s, Boeing would become a major supplier of pursuit airplanes to the US Army Air Corps (renamed from the US Army Air Service in July 1926), as well as the US Navy and Marine Corps. However, the story of these airplanes are outside the scope of this book.

**Above:** A head-on view of the second production Thomas-Morse MB-3, A.S.63332 in its original form. Note pitot on interplane strut. This particular MB-3 was sent to McCook Field as serial number P164. The airplane was later transferred to Boeing for use as a template. Via Colin Owers

**Above:** MB-3 A.S.63332, McCook P-164, presumably photographed from the roof of a hangar. The compactness of the MB-3 is clearly visible. Via Colin Owers

**Above:** A.S.63336 was the sixth production MB-3. It was subsequently transferred to McCook Field as P197. Of note is that the MB-3 still has the prototype fin/rudder arrangement. McCook photo 9974 via Colin Owers

**Left:** An up-close view of the MB-3A machine gun synchronising equipment. McCook photo 19254 via Colin Owers

**Above:** MB-3A A.S.68238 went to McCook Field as P260, being fitted with 'speed wings', consisting of a new airfoil and four ailerons. Via Colin Owers

**Above:** MB-3A A.S.68286 – marked as 68586 in error – was tested at McCook Field as P274, being equipped with faired lights into wing tips for night landings. Via Colin Owers

## In USAAS/USAAC Service

Following the end of the Great War (as it was commonly referred to at the time), the USAAS Pursuit squadrons were reorganized. On August 22, 1919, the 1st Pursuit Group (PG) was formed at Selfridge Field, Detroit, Michigan. The 1st PG consisted of four Pursuit Squadrons (PS); the 147th PS (later renamed 17th PS), which had been reconstituted on April 27, 1919, the 27th PS, reconstituted on April 28, 1919, the 94th PS, reconstituted on June 27, 1919, and the 95th PS, reconstituted on August 12, 1919. Initially, Curtiss JN-4s were flown before the 1st PG moved to Kelly Field outside San Antonio, Texas, on August 31, 1919. Having arrived in Texas, SE 5as, SPAD XIIIs and Fokker D.VIIs were delivered to the 1st

**Above:** An MB-3A makes interesting size comparison with one of the two Atlantic (Fokker) T-2s acquired by the Air Service, and the Verville-Sperry R-3 monoplane racer. McCook Field photo 16887 via Colin Owers

PG. Apart from forming the USAAS's first tactical Pursuit Group, one of its squadrons served as a conversion unit for pilots arriving from various flight schools. On June 30, 1921, the 1st PG transferred from Kelly Field to Ellington Field.

Prior to this, the first Curtiss-built Orenco D had arrived at Kelly Field in late November 1920. Initially, the Orenco D was considered an excellent pursuit aircraft. In the event, more than 50 Orenco Ds were delivered to the 1st PG. However, it was soon discovered that quality control during manufacture was sub-standard, resulting in a number of complaints. Following a fatal accident in 1921, the pilot's trust in the Orenco D evaporated, with the type being withdrawn soon afterwards.

With production of the MB-3 progressing well, it was said that 25 MB-3s were to be delivered to the 1st PG in April 1921. However, as described earlier, Lieutenant Macready's accident in an early production MB-3 resulted in delivery being delayed until early 1922. As a result of the problems with the Orenco D and non-delivery of the MB-3, the SE5as, SPAD XIIIs, and Fokker D.VIIs had to remain as first-line equipment until 1922.

The 1st PG pilots seemed fairly pessimistic about the home-grown Orenco D and Thomas-Morse MB-3, instead favoring the 200 h.p. Fokker D.VII and the SPAD XIII powered by the 180 h.p. Wright E engine. About 20 SE 5as were also serviceable, with only half that number being required for pilot training. Another SE 5a was flown by the Commanding Officer of the 1st PG, Major Carl Spaatz.

In March 1922, the first MB-3s finally arrived at Ellington Field. However, numerous problems with the new pursuits were soon reported: "Due to faulty construction of these planes, the time required for maintenance and necessary repairs and alterations to keep these planes in serviceable condition is excessive. It is the unanimous opinion of the pilots of the First Group (Pursuit) that the Thomas Morse MB-3 plane lacks may (sic) of the essential qualities required in a first class pursuit plane."

By May 25, 1922, 36 MB-3s had been delivered, with 20 being serviceable. Orders were received from the Chief of the Engineering Division at McCook Field to identify, as quickly as possible, the various modifications recommended for inclusion in the Boeing-built MB-3As. Intensive flight trials resulted in the majority of the MB-3s suffering structural damage, from bowed wing struts, leaky fuel tanks, cracked bulkheads and loose stabilizer fittings. Remarkably, following the conclusion of the flight trials, only one MB-3, A.S.63352, remained serviceable.

On June 22, 1922, flight training at Ellington Field ended, with the 1st PG being transferred back to Selfridge Field.

Meanwhile, the MB-3A delivery schedule called for fifteen aircraft (A.S.68340 to 68364) to be delivered to Mather Field for distribution among domestic USAAS units. Another fifty-one MB-3As (A.S.68365 to 68415) would also be shipped to Maher Field, and delivered overseas. These units included the 3rd Pursuit Squadron (reformed on January 25, 1923), based in the Philippines, and the 6th Pursuit Squadron based at Wheeler Field in Hawaii. The latter squadron was joined on May 1, 1923 by the 19th Pursuit Squadron. The 24th Pursuit Squadron

**Above:** MB-3A A.S.68237 seen at McCook Field. The airplane lacks the side radiators but is carrying a 50-lb bomb beneath the fuselage. Via Colin Owers

**Above:** he second production Thomas-Morse MB-3, A.S.63332 later went to McCook Field as P164. Note legend near cockpit in white: GAS CAP. 46 GALS. OIL 16 QRTS. It was later converted to MB-3A standard. Via Colin Owers

**This Page & Facing Page:** Boeing MB-3A A.S.68286 - erroneously painted as 68586 – McCook Field P274, was experimentally fitted with a ski undercarriage and a four-blade propeller. Note the bomb rack beneath the fuselage, underslung radiator, and lights in wingtips. The speed and maneuverability penalty with all of this added equipment must have been considerable. McCook Photos 20227, 20224, 20226 & 20225 via Colin Owers.

(redesignated as such on January 25, 1923), based at France Field in the Canal Zone from April 1922, also received MB-3As. The final twenty-one MB-3As (A.S.68416 to 68436) were to be transferred to Kelly Field's 43rd School Squadron.

MB-3As were also used by the Air Service Tactical School (ASTS) and the Air Service Technical Training School (ASTTS).

In the mid-1920s, several MB-3As were rebuilt by the Fairfield Air Intermediate Depot (F.A.I.D). These aircraft were sent to Kelly Field near San Antonio, TX, where they were used for advanced training by the 43rd School Squadron. Here they joined obsolete MB-3s modified as advanced trainers and redesignated as MB-3Ms. Some MB-3Ms were still in service in 1929.

## In USMC Service

Originally built to an Army contract (serial numbers A.S.64374-64384), 11 MB-3s were purchased by the US Navy, but did not serve with the Navy; instead they were transferred to the US Marine Corps. The original contract, dated May 16, 1921, covered ten MB-3s and two MB-7s. When one of the MB-7s was

destroyed during the 1921 Pulitzer Air Race, it was replaced by an eleventh MB-3, ordered on December 2, 1921. This contract was a "rare example of joint Army-Navy administrative cooperation".

The final MB-3 was delivered in February 1922. Upon arrival at Gerstner Field, Louisiana, it was discovered that the MB-3s engines and armament were in poor condition, requiring refurbishment. Issued with the serial numbers A-6060 to A-6070, the MB-3s were shipped to Norfolk, before being transferred to Quantico. The MB-3s equipped 'F' Flight of the Third Air Squadron, providing a nucleus for developing pursuit aircraft tactics.

One USMC pilot, Captain Ford O. 'Tex' Rogers, had to bale out of s/n A-6062 when manoeuvring in rough weather. Rogers' verdict was that the MB-3 was "fast, tricky and as tiring as hell to fly (and that) they were somewhat prone to ground loop." The remaining MB-3s were consigned to storage in July 1922, having served for less than six months. However, it was not until November 1923 that the aircraft were sold back to the USAAS.

## The MB-3 As An Air Racer

During the early 1920s, USAAS aircraft regularly participated in air races. Air racing caught the imagination of the American general public, with such events usually drawing huge crowds. The

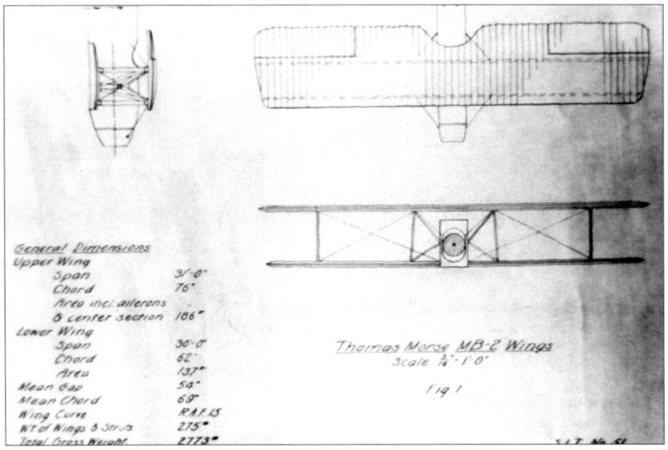


**Above:** A contemporary three view drawing of the MB-2. Via Colin Owers

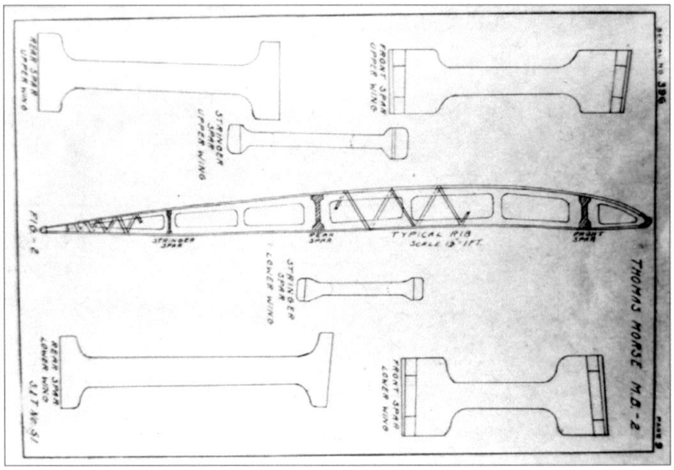


**Above:** A rib drawing for the MB-2. Via Colin Owers

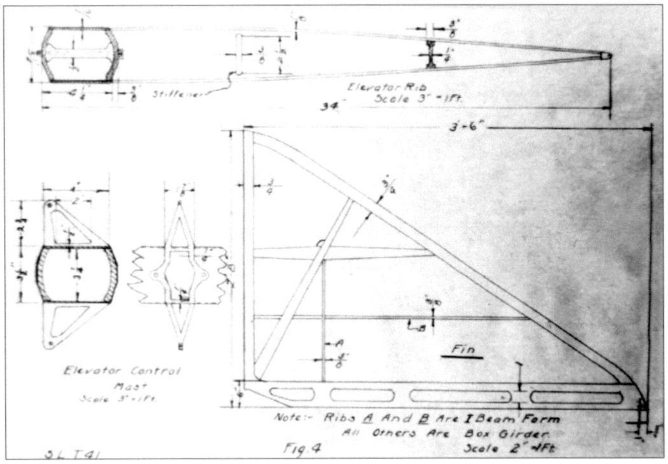


**Above:** Another rib drawing for the MB-2. Via Colin Owers

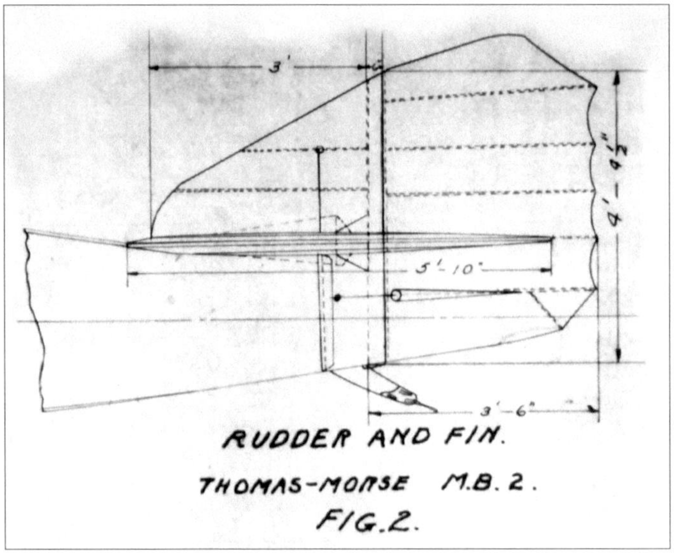


**Above:** MB-2 fin and rudder drawing. Via Colin Owers

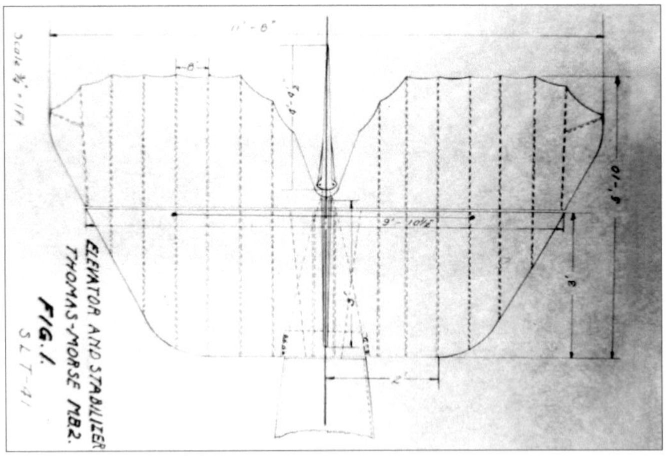


**Above:** MB-2 Elevator and stabilizer drawing. Via Colin Owers

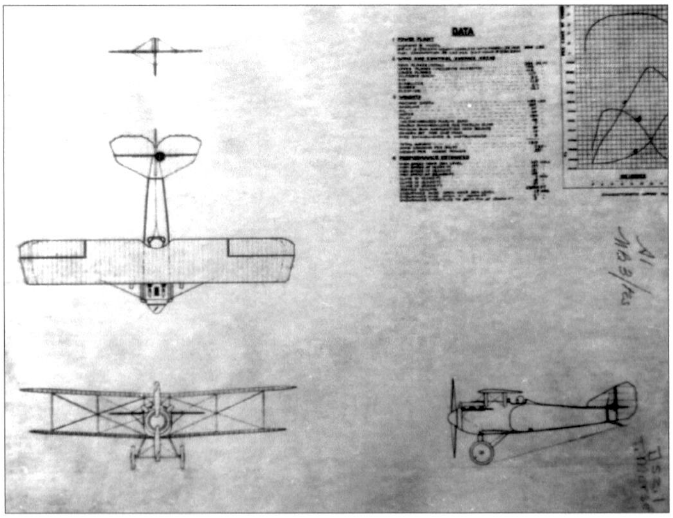


**Above:** A basic general three-view drawing of the MB-3. Via Colin Owers

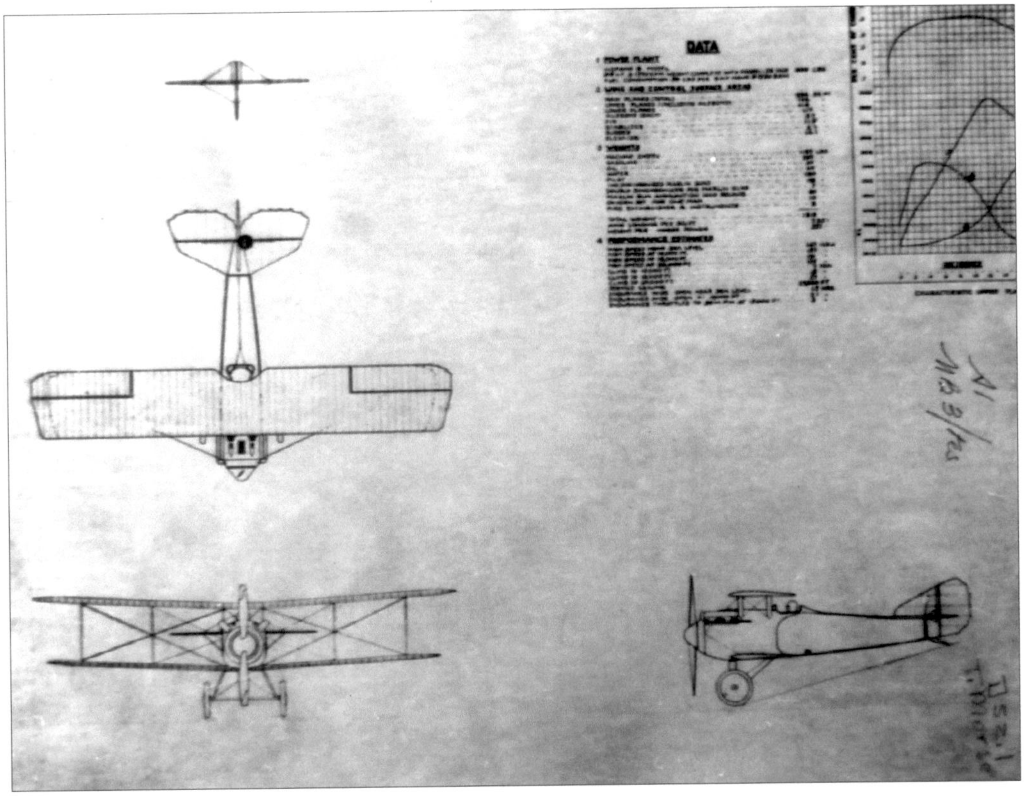


**Right:** A basic general three-view drawing of the MB-3. Via Colin Owers

**Below:** A contemporary drawing of the Boeing MB-3A. Via Colin Owers

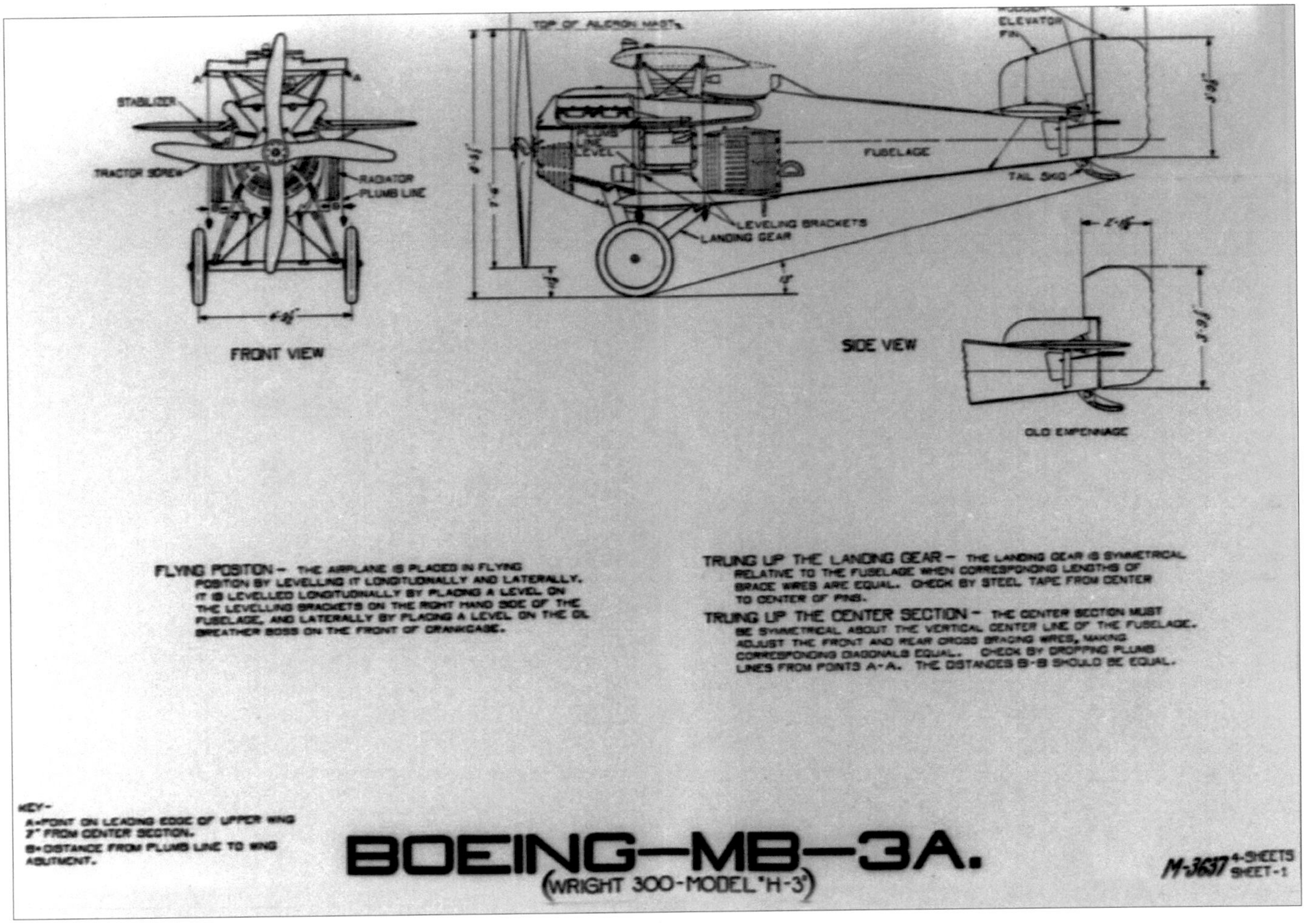

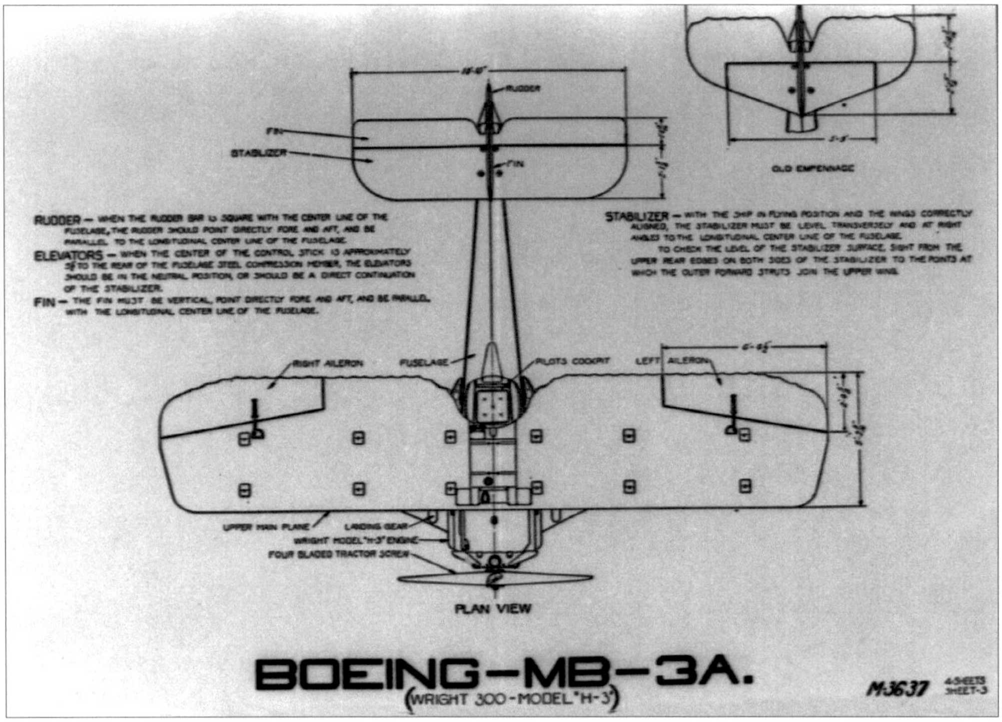


**Left:** An upper view of the Boeing MB-3A. Via Colin Owers

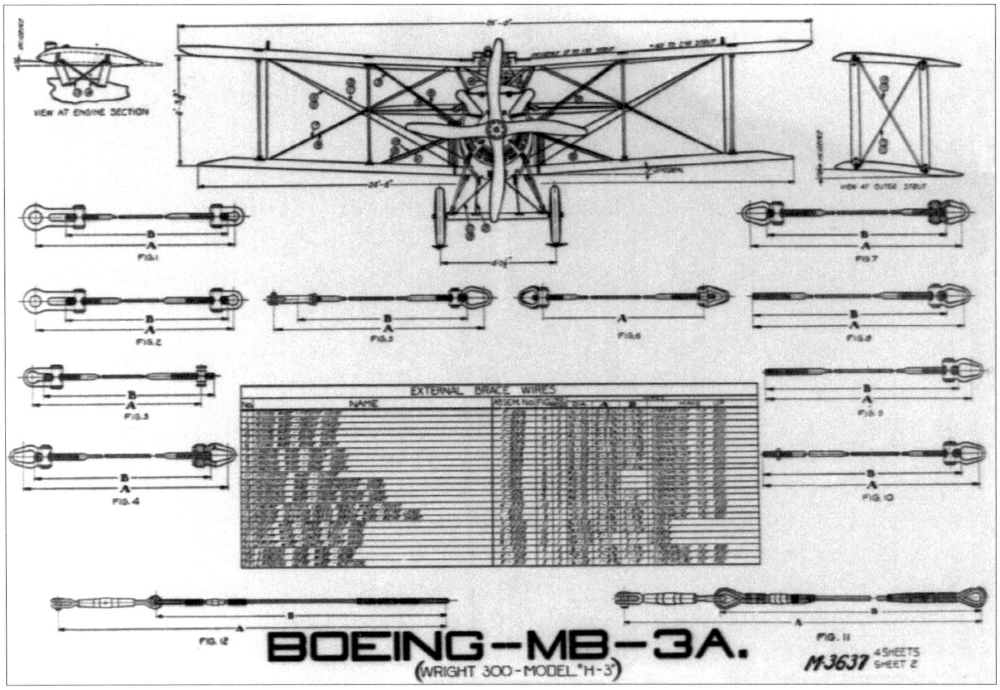


**Left:** A full frontal view of the Boeing MB-3A. Via Colin Owers

participating aircraft represented the forefront of aeronautical development, both with regards to airframe and engine. As a result, much money and resources were invested in developing air racing aircraft by various manufacturers across the USA, with both the USAAS and US Navy showing a keen interest in such aircraft. The first air race in which a Thomas-Morse aircraft took part was the 1920 Pulitzer Trophy Race, held on Thanksgiving Day at Mitchel Field, Long Island. The race was named after Hungarian-born Joseph Pulitzer (1847-1911), who after immigrating to the USA in 1864, made his fortune in publishing. Late in his life, Pulitzer developed an interest in aviation.

In 1916, five years after the death of the father, Pulitzer's three sons revealed plans to stage a Coast-to-Coast Continental air race. The race was to take place following the end of the Great War, and then become an annual event. Such a race was held in the summer of 1919. However, out of the 63 contestants,

**Above:** A rare air-to-air view of three USMC MB-3s. Via Colin Owers

only eight completed the cross-country return flight. Tragically, nine aviators died during the race. As a result, the cross-country race was not repeated.

For the 1920 Pulitzer Trophy Race, four laps of a triangular 33-mile race track was to be completed. Out of the 45 participants, two were MB-3s. These were the second and third prototypes respectively. The first of these was flown to second place (the winner being the Verville-Packard VCP dedicated air racer) by Captain Harold Evans Hartney at an average speed of 148 mph. Hartney's MB-3 (the second prototype, A.S.40093, racing number 41) was a standard aircraft, not having been modified prior to the race. Flying the second MB-3 (A.S.40094, racing number 43), 2nd Lieutenant Leigh Wade had to abort on the second lap due to a technical malfunction. Having an unmodified MB-3 ending up in second place spoke volumes for the MB-3s performance, and, in particular, its speed. The fact that an MB-3 had ended up in second place could possibly have influenced the decision to order the type into series production.

The 1922 event did take place between October 7 and 14 at Selfridge Field outside Detroit. There were 15 USAAS and US Navy contestants, the War Department having apparently realized that military participation would be worth both the cost and effort. Although half-a-dozen MB-3As had been entered in the Pulitzer Air Race, the participation of the MB-3As were eventually reduced to that of the John L. Mitchell Trophy Race. It was thought that the MB-3As would stand little or no chance against purpose-built air racers. The latter included the all-metal Thomas-Morse R-5 (also known as the TM-22), which was powered by a 600 h.p. Packard engine. Nevertheless, two quickly modified MB-3As, racing numbers 61 and 62, were prepared to take part in the race. The MB-3As were to be flown by Lt Thomas K. Matthews and Lt George P. Tuortellot, were withdrawn before the race was to begin. In the end, the R-5s ended up at the lower end of the field, finishing in tenth and eleventh places. Disassembled and transferred to McCook Field, one of the R-5s were flown intermittently before being grounded in mid-1923. Consigned to storage, both R-5s were allotted for static fatigue tests.

Instead of participating in the National Air Races (an all-embracing name for the different kinds of events involved), the John L. Mitchell Trophy Race was created by General Billy Mitchell. This was in honor of the memory of his younger brother, who had been killed in a flying accident in France on May 26, 1918. General Mitchell stipulated that only pilots and aircraft of the 1st Pursuit Group were allowed to compete in the John L. Mitchell Trophy Race. The six MB-3As originally destined to take part in the Pulitzer Air Race were assigned to the Mitchell Trophy Race instead. In order for any speed records attained over the 50 km (31 mile) triangular course to be registered with the

**Above:** The MB-3 prototype seen during static tests. Via Colin Owers

**Above:** The MB-3 undercarriage was part of the static test program. Via Colin Owers

**Above:** On August 18, 1919, the MB-3 undercarriage cracked during the static test program. Via Colin Owers

**Above:** Another view of the destroyed undercarriage. Via Colin Owers

**Above:** Another photographic evidence during the static test program. Via Colin Owers

Federation Aeronautique International (FAI), the metric system was used. Recalculating miles into kilometres was seen as difficult.

The six pilots were; Capt. O.W. Broberg (racing number 55), Capt. H.M. Elmendorf (53), Capt. A. Guidera (51), Lt B.K. McBride (52), Lt D.F. Stace (54) and Lt J.D. Summers (56). The youngest of the six pilots was 22-year old Donald Stace, who had accumulated less than 100 hours in the air. However, Stace was the unit's Assistant Group Engineering Officer, serving with the Service Squadron. Having a pilot from the Service Squadron beat the pilots from the tactical squadrons was seen as a huge incentive. In order to increase the chances of victory, Stace's MB-3A, A.S.68257, underwent modification work to increase its speed. This included removing all appurtenances, such as radiator shutters and upper wing compass, fairing in the wing struts, as well as literally polishing the pursuit until it shone. The

**Above:** The second MB-3 prototype, A.S.40093, presumably at McCook Field. Note the old type of national insignia beneath the lower right wing. Via Colin Owers

**Above:** An MB-3 seen fitted with an abnormally long pitot tube. Via Colin Owers

night before the race, the MB-3A was varnished with the fuselage footstep being faired over. Stace also managed to fly the course beforehand in a SPAD XIII.

At 1135 hours on October 14, the six MB-3As were lined up for take-off, watched by 25,000 eager spectators. Stace was first off, followed in quick succession by McBride, Summers, Broberg, Elmendorf and Guidera. The six MB-3As passed across the starting pylon in tight formation, with Stace in the lead, his careful preparations appearing to work as intended. Incredibly, Stace gained one minute for each of the three laps, with Elmendorf losing one minute. On the third lap, Summers had to abort, landing back at Selfridge Field. Despite having his leather helmet blown away, Stace was first across the finishing line.

His average speed was 148 mph, having

**Above:** The second production MB-3, A.S.63332, was tested at McCook Field as P164. Note that the photo appears to have been retouched. Via Colin Owers

**Above:** Another view of MB-3 A.S.63332 at McCook Field. Via Colin Owers

remained in the air for 50 minutes and 25.73 seconds. Guidera was the runner-up at 136,1 mph (54 minutes and 48.32 seconds). Broberg came in third at 135,3 mph (55 minutes and 7.87 seconds). At fourth place was McBride at 134,6 mph (55 minutes and 23.11 seconds) with Elmendorf occupying fifth place at 124,7 mph (59 minutes and 48.77 seconds). Having been forced to abort on the third lap, Summers must have been hugely disappointed as his average speed on the two laps was 137 mph. Had he been able to continue, Summers could have occupied second place in the race.

This first John L. Mitchell Trophy Race was considered a huge success, becoming known as

**Right:** The fuselage frame of the first MB-3 prototype undergoing static tests. Via Colin Owers

**Right:** Before the advent of computers, airframe static tests had to be performed using 'old school' methods. Via Colin Owers

**Right:** Another view of MB-3 A.S.40092 during the static test program. Via Colin Owers

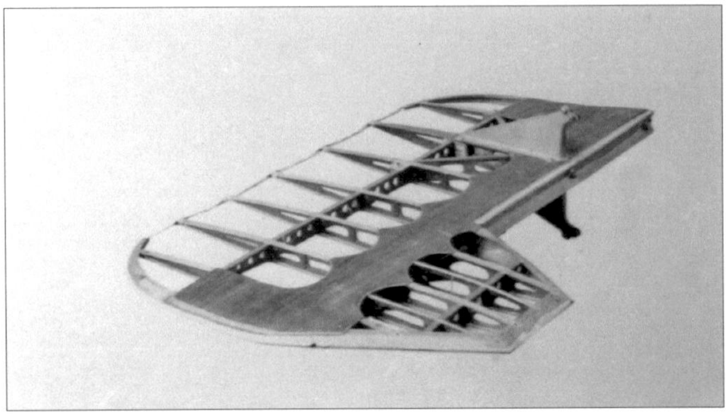

**Above:** Detail view of an uncovered elevator. Via Colin Owers

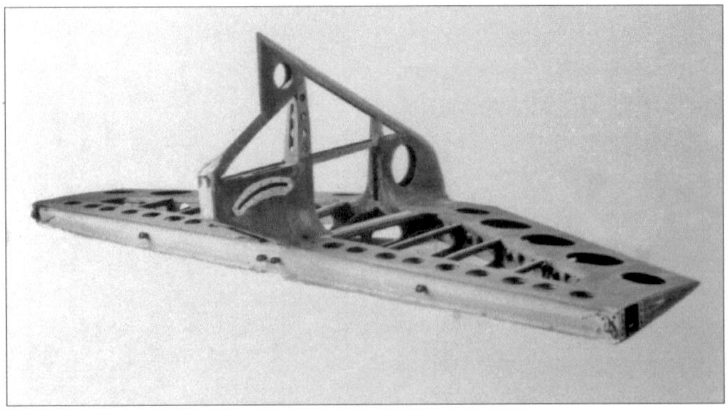

**Above:** An excellent view of the stabilizer/fin wooden structure. Via Colin Owers

**Left:** As the sign states, the MB-3 wings failed the static sand bag test on April 9, 1919. Via Colin Owers

**Below:** An excellent photo showing MB-3 prototype turned up-side-down prior to wing loading static tests. Note that the lower surfaces of the wings (or, upper, given that the airplane is mounted up-side-down, are uncovered). Via Colin Owers

**Right:** A worker holding up a canvas screen in front of the MB-3 prototype. Note the stabilizer in the background. Via Colin Owers

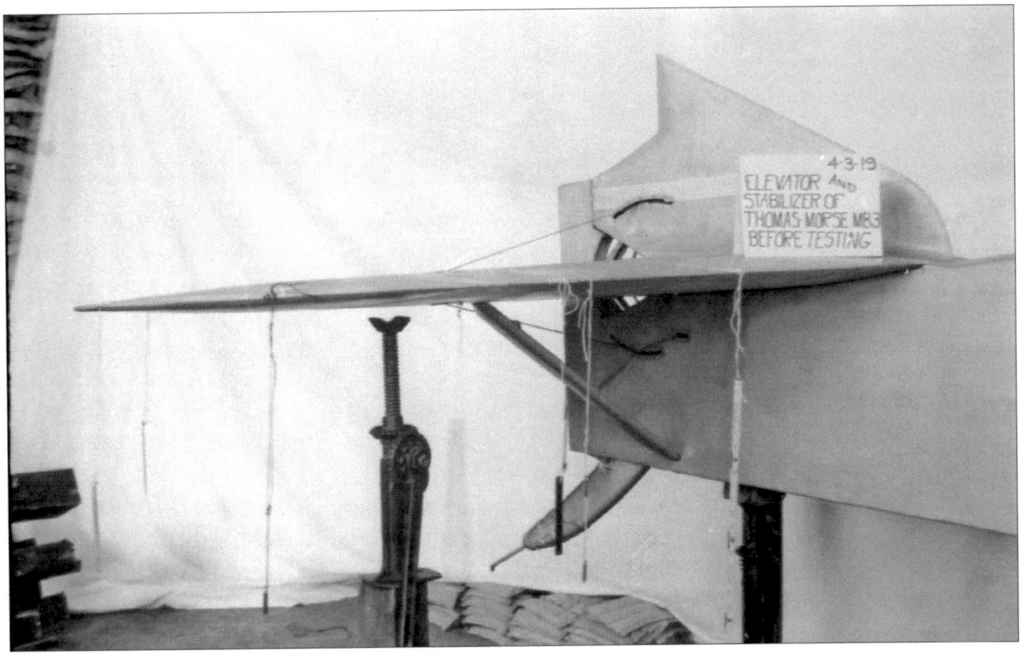


**Right:** Thankfully, most photos taken during the MB-3 static tests include a sign showing the date, and the purpose of the particular test. Here, the stabilizer is seen on April 3, 1919, prior to commencement of the sand bag load test. Via Colin Owers

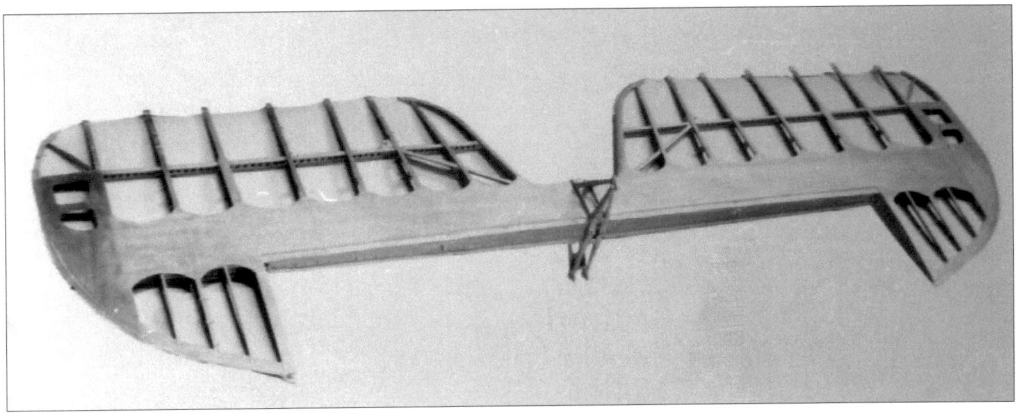

**Right:** A detail photo of the elevator. Note that the lower surfaces have been fabric-covered. Via Colin Owers

**Above:** A rare photo showing the first MB-3 prototype fully-assembled. Presumably, this photo comes from a contemporary publication. Via Colin Owers

the 'Thomas-Morse Race'. Apart from securing the trophy, Stace also received $ 400 from car manufacturer Dodge to be spent on a shiny new Dodge automobile. The MB-3A flown to victory by Lt Stace crashed on January 26, 1923, during a gunnery exercise. The pilot, Lt Schulz, was killed.

The 1923 National Air Races was due to take place at Lambert Field outside St. Louis, Missouri, on October 4, 5 and 6. For some reason, the itinerary initially did not list the John L. Mitchell Trophy Race among the events. Nevertheless, five MB-3A pilots – and their respective alternates – were announced on August 20 as participants; 1Lt Thomas W. Blackburn (alternate 1Lt Leland C. Hurd), 1Lt Thomas K. Matthews (1Lt Hobart R. Yeager), 1Lt George P. Tourtellot (1Lt Arthur J. Liggett), 1Lt J. Thad Johnson (1Lt Edward M. Haight), and 1Lt Russell L. Meredith (1Lt Louis C. Simon).

The participant's list was amended in late September to add a sixth pilot, Capt Vincent B. Dixon, with Capt Burt E. 'Buck' Skeel, while removing 1Lt Meredith. The sixth alternate would be Frank O'D Hunter. By including a sixth aircraft, this would emulate the 1922 event.

A mere six weeks prior to the National Air Races scheduled dates, the John L. Mitchell Trophy Race was eventually included as event 8a.

Despite poor weather, upwards of 100,000 spectators gathered at Lambert Field.

For the John L. Mitchell Trophy Race, MB-3As of the 1st Pursuit Group flew four laps over a 124 mile (50 km) course, the same as the 1922 race. The winning pilot would be given the chance to fly the most modern and high-powered air racer in the following year's event. Taking off at 1300 hours, the six pilots flew at extremely low altitude, further enhancing the spectator's desire for action. At the start of the third lap, problems with the fuel lines forced Matthews to land prematurely. Just before crossing the finishing line, the engine of Tourtellot's MB-3A quit due to lack of fuel. Putting the aircraft into a dive, Tourtellot hit the ground hard. Incredibly, his MB-3A bounced upwards making it across the finishing line. However, Tourtellot had been beaten by B E Skeel. Having taken off last, Skeel secured first place at an average speed of 146.45 mph (50 minutes and 54.95 seconds). Tourtellot's speed averaged 143.21 mph (52 minutes

**Above:** Compared to the photo above (page 46), the wings have been fabric-covered. Via Colin Owers

and 4.24 seconds). Ending up in third place was T W Blackburn at141 mph (52 minutes and 51.07 seconds). Competition between J T Johnson and V B Dixon was intense, with Johnson securing fourth place at 139.20 mph (53 minutes and 34.18 seconds), beating Dixon by a margin of just over six seconds. Dixon's average speed was 138.94 mph (53 minutes and 40.22 seconds).

The day of the Thomas-Morse MB-3A as an air racer had passed. Purpose-built air racers were to dominate the following year's events. Tragically, Bert Skeel would perish when his Curtiss R-6 crashed during the 1924 National Air Races.

**Ski-equipped MB-3A A.S.68286, erroneously marked as A.S.68586.**

**Above:** Another view from the wing failure static test on April 9, 1919. Via Colin Owers

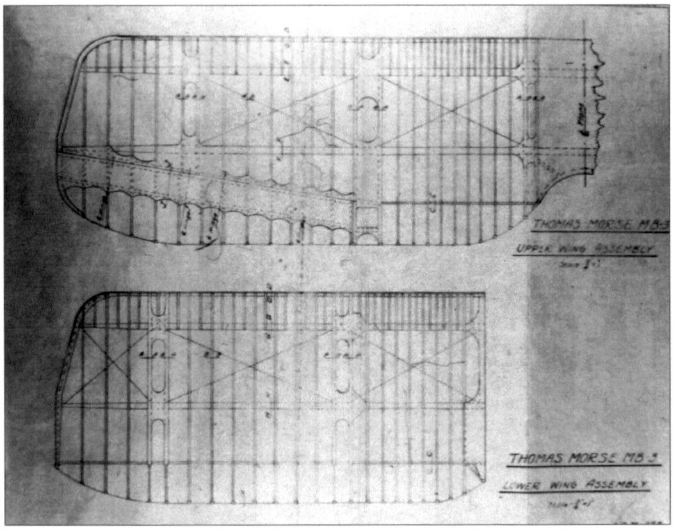


**Above:** A factory drawing showing the upper and lower wing assembly. Via Colin Owers

**Facing Page, Above:** A USAS drawing of the MB-3. Via Colin Owers

**Facing Page, Below Left:** A factory drawing showing the MB-3 rudder assembly. Via Colin Owers

**Facing Page, Below Right:** A factory drawing showing the fin assembly assembly. Via Colin Owers

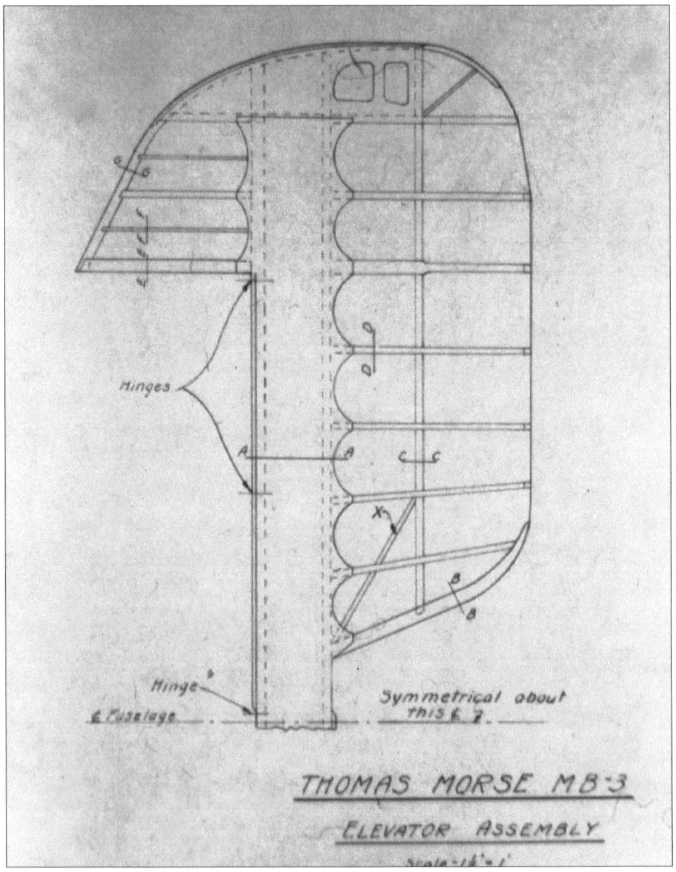


**Above:** A factory drawing showing the MB-3 elevator assembly. Via Colin Owers

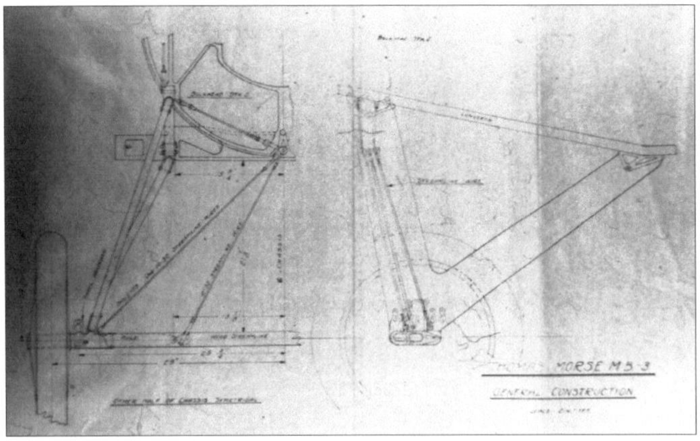


**Above:** A factory drawing showing the general construction of the undercarriage. Via Colin Owers

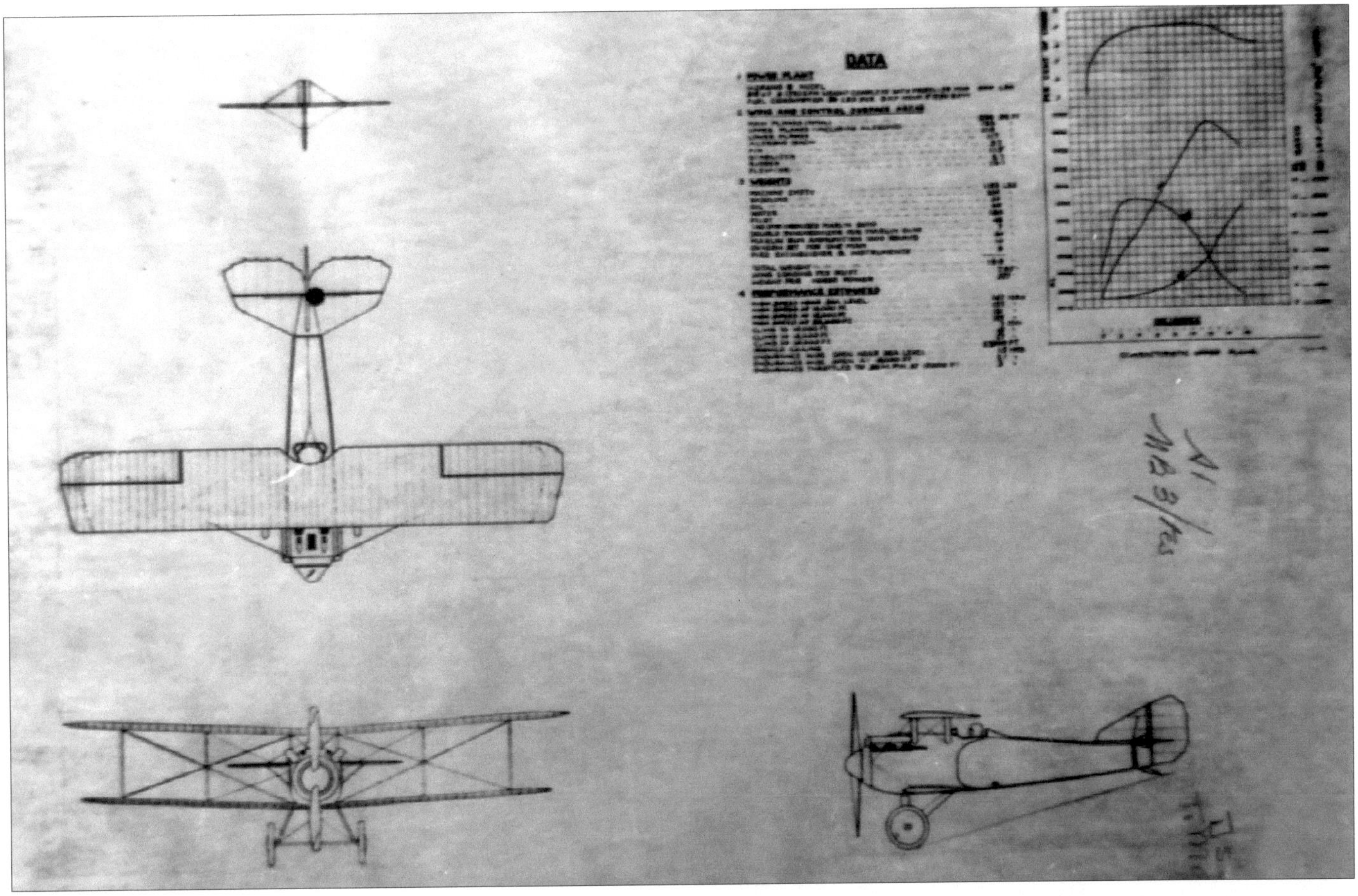
DATA

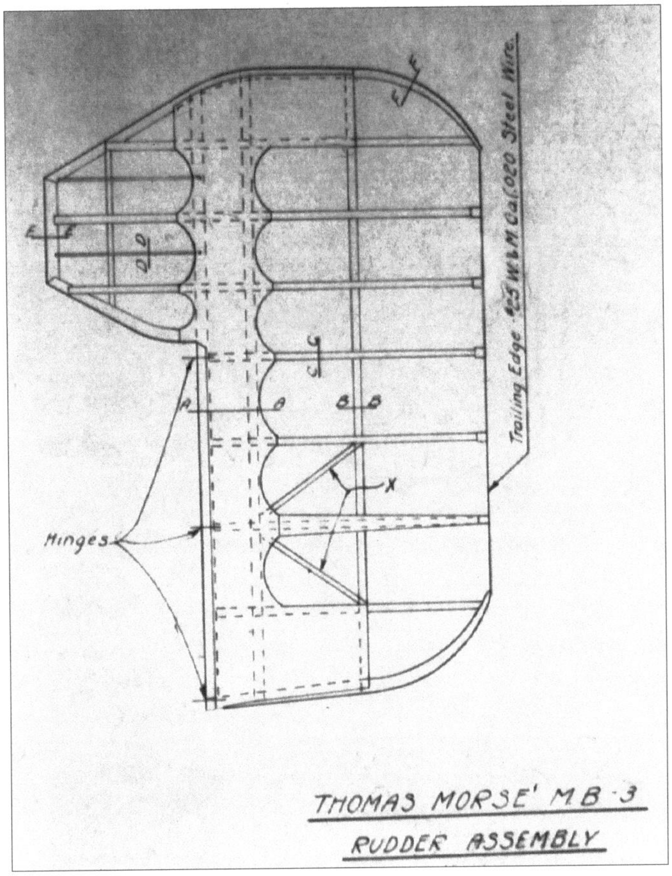
Trailing Edge
Hinges
THOMAS MORSE' MB-3
RUDDER ASSEMBLY

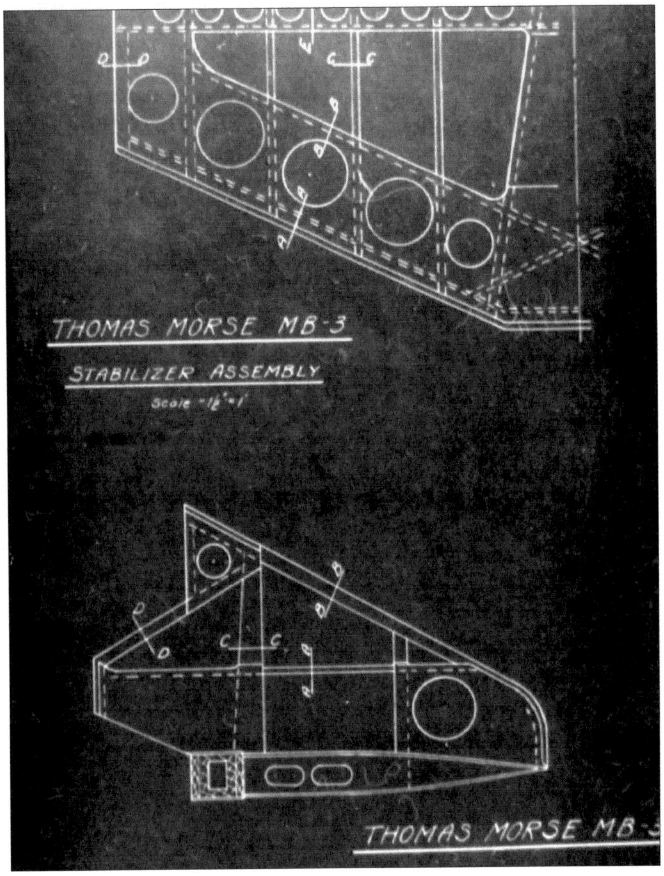
THOMAS MORSE MB-3
STABILIZER ASSEMBLY

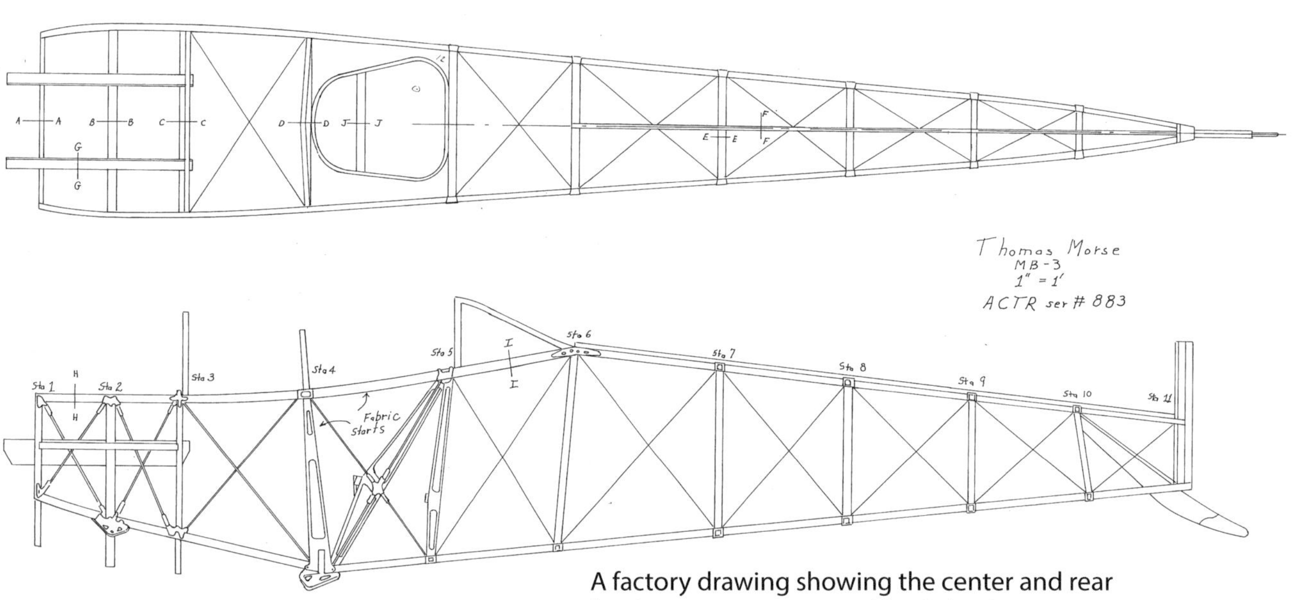


A factory drawing showing the center and rear fuselage structure. Traced by Colin Owers

**Above:** A close-up detail photo of an MB-3A. Note the underwing bomb rack. Via Colin Owers

**Above:** Female workers working on the upper wing of a Boeing MB-3A. Via Colin Owers

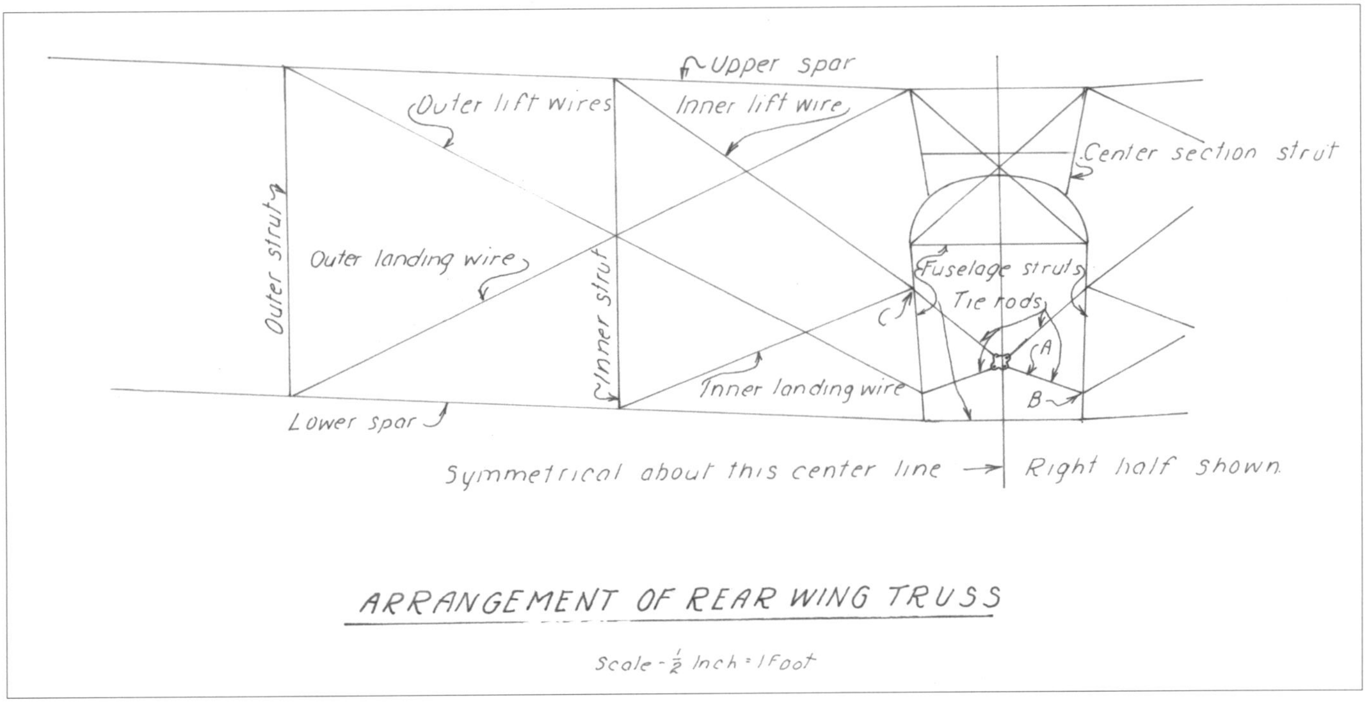


**Above:** A factory drawing showing the arrangement of the rear wing truss. Via Colin Owers

**Above:** A close-up of the destroyed MB-7. Via Colin Owers

**Facing Pagee:** The MB-7 was completely destroyed on impact. Via Colin Owers

**Right:** The MB-7 soon after meeting its end. Via Colin Owers

**Above:** An unidentified MB-3 was also consumed during the crash tests. Via Colin Owers

**Facing Page, Above:** The MB-7 slams into the wall. Via Colin Owers

**Facing Page, Below:** An MB-3 with wings and fuselage fabric covering removed prior to a destructive test. Via Colin Owers

**Above:** An MB-3 about to be sacrificed for the sake of improved aviation safety. Via Colin Owers

**Above:** An MB-3A at an unknown location. Note the Martin NBS-1 in the background. Via Colin Owers

**Facing Page, Below:** An MB-3 used in the movie *Wings*. For the movie the MB-3s were painted up in a multi-color splotchy camouflage vaguely resembling the French five-color camouflage system – but even on the undersides of the wings in the movie, whereas the originals had one color. The fins and rudders of the MB-3's were crudely painted to attempt to resemble the scalloped trailing rudder edge and fin profile of the Spads, and "SPAD XIII" stenciling was applied to the rudder stripes. The "Squadron insignia" of a bald eagle pouncing on a dachshund with an Iron Cross on its belly was fictitious but closely resembled the actual wartime insignia of the 24th Aero Sqdn – which flew Salmsons. via Greg VanWyngarden.

**Above:** An MB-3A performing in front of film cameras. It is unclear if this photo was taken during the shooting of *Wings*. Via Colin Owers

**Above:** An MB-3 used in the movie *Wings*. Note the rudder, rudder, and elevators were painted as scalloped to resemble a Spad. *Wings* was the first WWI air war movie produced by Hollywood, directed by former Lafayette Flying Corps pilot William Wellman, and was a massive, gigantic production famous for being the very first movie to win the Academy Award for Best Picture. via Greg VanWyngarden.

**Above:** MB-3s used in the movie *Wings* housed in a hangar between takes. via Greg VanWyngarden.

**Above:** MB-3s lined up on the set of the movie *Wings*. The tent hangars resembled those of WWI. via Greg VanWyngarden.

**Above & Below:** MB-3s in scenes from the movie *Wings*. via Greg VanWyngarden.

**Above & Facing Page, Above:** MB-3s from the movie *Wings* in their fictitious squadron insignia. via Greg VanWyngarden.
**Facing Page, Below:** One of the two MB-3s deliberately crashed for the movie *Wings*. Other, unplanned crashes occurred during filming. via Greg VanWyngarden.

# The Thomas-Morse MB-4

Designed in response to a US Post Office Department specification, the MB-4 two-seat biplane used major components of the MB-3 pursuit. Issued in June 1919, the specification called for ten aircraft capable of carrying 1,500 lb of air mail when flying on one engine. The aircraft was to be fitted with two or three Liberty or Hispano-Suiza engines. For its time, the MB-4 basic configuration was quite novel. The two Wright H engines, mounted in a tractor-pusher configuration in a centrally placed nacelle, were flanked on each side by modified MB-3 fuselages. The primary pilot sat in the port fuselage, with the starboard fuselage housing the second pilot/mechanic. The payload was located in the center nacelle. The three-bay wings were of unequal-span. The MB-4 was the first tractor-pusher aircraft constructed in the USA. Completed in early 1920, the MB-4 was test flown at Ithaca on a few occasions in February. The MB-4 exhibited quite a few substandard features, including no possibility for the two pilots to communicate when airborne, the engines causing severe airframe vibration, an inadequate fuel system, etc. Take-offs were usually asymmetric, with one of the fuselages lifting off the ground before the other. The Thomas-Morse factory superintendent Jerome Fried held a rather pessimistic opinion of the MB-4, calling it the 'worst thing on wings.' Post Office Department officials were present for a March 1 test flight. In the event, the Post Office Department decided against ordering the MB-4. The prototype was scrapped in 1921. However, the USAAS briefly considered a bomber development of the basic MB-4. Three MB-4s, serial numbers A.S.64306 and 64373/64374 were ordered for evaluation. The first of these arrived at McCook Field on September 23, 1921. Fitted with two 300 h.p. Wright H engines (serial numbers 13506 and 13508), the MB-4 was assigned the McCook Field serial number P172. Unfortunately, the outcome of these tests remain unclear.

### MB-4 Technical Data and Performance Characteristics

Wingspan: 45 ft 6 In (13,87 m) (Upper) 41 ft (12,5 m) (Lower)
Length: 25 ft 4 ¾ in (7,71 m)
Height: 7 ft 10 in (2,39 m)
Empty weight: 3,554 lb 1,613 kg)
Gross weight: 5,564 lb (2,526 kg)
Maximum Speed: 140 mph (225 km/h)
Climb to 10,000 ft (3,048 m): ten minutes
Range: 600 miles (965,4 km)

# The Thomas-Morse MB-6

A dedicated air racing variant of the MB-3, designated MB-6, was to represent the USAAS in the 1921 Pulitzer Air Race, due to be held in Detroit, Michigan, in November that year. On 24 May 1921, a $ 48,000 contract was placed with Thomas-Morse for three MB-6s, serials 68537, 68538 and 68539. According to the contract, the MB-6 was to have a top speed of 175 mph, and a landing speed of 75 mph. The MB-6 was basically an MB-3 with reduced wingspan, having single-bay wings. One other difference was that the single radiator was fitted in front of the engine. The USAAS was to supply Wright-built Hispano-Suiza H-2 engines, propellers and instruments. However, in July 1921, the War Department decided that participation in the forthcoming air race would be too expensive, thus withdrawing military support for the MB-6 as well as the Navy's MB-7. Thinking that the lack of military aircraft would reduce public interest in the event, Detroit officials withdrew their support. The following month, a new venue was found in Omaha, Nebraska. Nevertheless, work on the three MB-6s continued.

According to the original delivery schedule, the MB-6s were due to be delivered by August 23. However, the delivery date had to be amended twice, with the three MB-6s not being handed over until September 20. Orders soon arrived to use the second MB-6, A.S.68538, for static weight tests. Flight testing at McCook Field was to commence when the static tests had been completed. MB-6 68537 was flown for the first time on October 21, with Lieutenant. John A. Macready at the controls. During the fifteen-minute flight, an average speed of 110 mph was attained, with one high speed run of 182 mph. Macready stated that the MB-6 was "tricky, but not dangerous." A second flight, this time with Captain Corliss C. Moseley at the controls, took place on October 23. Two days later, the MB-6 was destroyed in a landing accident. The cause was found

**Above:** Thomas-Morse MB-6 racer. The MB-3 ancestry is evident. Via Colin Owers

to be a faulty fuel pump. Moseley was lucky to escape serious injury. The third, unassembled MB-6, A.S.68539, had been shipped to Omaha by Thomas-Morse at the company's own expense.

Flown by Lieutenant Macready, the MB-6 (assigned racing number 2) ended third with an average speed of 160,83 mph. Having crossed the finishing line, the Wright H-2 engine stopped. During the race, the engine had been run at full throttle, overheating slightly. Macready put the MB-3 into a dive to avoid stalling, attempting a downwind landing. The engine then decided to start working once again, with Macready making a normal landing.

The MB-6 was then disassembled, and returned to McCook Field. Although reassembled and redesignated as the R-2, it appears that the aircraft was not flown again. Consigned to storage, the R-2 was removed on October 31, 1924, to suffer the ignomy of being used for crash tests. Placed on a track, the R-2 was then crashed into a concrete wall, completely destroying the aircraft.

### MB-6 Technical Data and Performance Characteristics

Wingspan: 19 ft 5,79 m)
Length: 18 ft 6 in (5,64 m)
Height: 7 ft 10 in (2,39 m)
Weight: 2,023 lb 918 kg)
Wing Area: 157.11 sq ft (47,9 $m^2$)

# The Thomas-Morse MB-7

The MB-7 was custom-built for the US Navy for participation in the 1921 Pulitzer Air Race. It was part of a May 16, 1921 contract, covering ten MB-3s and two MB-7s to be supplied to the US Navy. USMC The aircraft were due to be delivered in November 1921. Although the fuselage and tail surfaces remained similar to that of the MB-3 and MB-6, the MB-7 was a high-wing monoplane. The wing was of novel design, known as the Alula wing. Developed in Great Britain by the Commercial Aeroplane Wing Syndicate, Ltd, the leading edge of the Alula wing tapered off sharply towards the wing tips. This would, it was thought, reduce the span-wise flow and the resulting tip losses. For the MB-7, the Alula wing was further modified, having a thicker airfoil as well as reduced wing taper to increase aileron effectiveness. When viewed from straight ahead, the MB-7s wing had the appearance of a humped gull wing. Thomas-Morse test pilot Paul D Wilson performed initial taxiing trials.

On October 24, 1921, Harold Evans Hartney performed the MB-7s maiden flight at Ithaca. By this time, Hartney had been promoted to Colonel. The following week, the MB-7, A.S.64373, was shipped to Omaha. However, due to the War Department cancelling military participation on the grounds of high cost, the MB-7 was sponsored by Thomas-Morse.

Following arrival at Omaha, the MB-7 was allotted racing number 2. However, fuel pump problems proved so severe that Hartney did not manage to take-off until after all the other participants had landed. Unfortunately, the fuel pump failed yet again during the race, with Hartney crashing at the Campbell family farmstead. Hartney was thrown clear when his MB-7 impacted the ground, suffering a broken thigh bone and a fractured hip. Assisted by members of the Campbell family, Hartney told them to "Telephone my wife and tell her I've sprained my ancle".

The crashed MB-7 was soon at the mercy of 'souvenir hunters' who proceeded to strip the wreck of anything that could be carried away. To add insult to injury, one bystander, Fred Offertag, carelessly lit a cigarette, tossing the match near the remains of the MB-7. The wreck caught fire, being completely destroyed. When told about the demise of his aircraft, Hartney burst into tears.

The second MB-7, serial number A.S.64374, was completed in February 1922. Powered by a 400 h.p. Wright H-3 engine, it was otherwise identical to the first MB-7. In March 1922, the second MB-7 was shipped to Mitchel Field. Here it was flown by Capt. Francis P. Mulcahy, who later reminisced that "The MB-7 was continually a source of trouble due to the insufficient cooling of the engine at high power settings." Nevertheless, Mulcahy remembered that the MB-7 performed well, although the poor forward view made landings problematical.

Mulcahy flew the MB-7 in the 1922 Pulitzer Air Race, where it was allotted racing number 7. However, due to the engine overheating – despite an auxiliary radiator having been fitted - Mulcahy was forced to drop out after two laps, having averaged a speed of 145 mph on the first lap.

The MB-7 was then transferred to the USMC, with the serial number being changed to A-6071. In his book United States Navy and Marine Corps Fighters 1918-1962, Paul R Matt claims that the MB-7 was classified as a pursuit, despite not being armed, and used as a continuation trainer.

Another source states that the MB-7 was shipped to the Naval Aircraft Factory at Philadelpia, eventually being stricken on May 5, 1925.

### MB-7 Technical Data and Performance Characteristics

Wingspan: 24 ft (7,315 m)
Length: 18 ft 6 in (5,64 m)
Height: 7 ft 3 in (2,21 m)
Wing Area: 112 sq ft (34,14 m$^2$)
Gross Weight: 2,000 lb (908 kg)
Maximum Speed: 180 mph (290 km/h)

## Movie Stars

The visual legacy of the MB-3A is completely due to the 1927 movie *Wings*. Perhaps the last great of Hollywood's silent movies before the advent of 'talkies', *Wings* was a massively huge undertaking. The aerial combat sequences were created using almost the entire fleet of Air Service combat airplanes. The War Department made available some 220 aircraft (including Curtiss P-1s starring as German pursuits, DH 4s, Douglas O-2s, and Martin NBS-1 bombers), along with other equipment and personnel. However, the participating MB-3As are not playing themselves, but rather French SPAD and German Albatros D.III and Fokker D.VII pursuits! The MB-3As were modified to fit their roles as French and German pursuits. Apart from being adorned with French-style camouflage, the MB-3A 'SPADs' had their engine cowlings and machine gun installation modified to make them more similar to the original SPADs. Additionally, the fin was painted in such a way as to imitate a SPAD XIII fin. At this stage, the MB-3A was thoroughly obsolete as a pursuit, with several being utilized as advanced trainers. One might speculate that if actual filming would have begun in late 1927, no MB-3As would have been involved.

Two MB-3As obtained by Paramount were intentionally destroyed during filming. The respective aircraft met their end by being fired into the air by means of an air pressure gun. One of these scenes involved David Armstrong's (Arlen) 'Fokker D.VII' ending up against a building. The fin and rudder of this particular MB-3A had been modified to look like that of a Fokker D.VII. The second MB-3A was sent crashing into a lake. One (other?) MB-3A was modified as an Albatros D.III (including the removal of the inner wing struts), and set up in an inverted position in front of the cameras.

Remarkably, only one fatal crash occurred during the filming of *Wings*. This occurred in November 1926, MB-3A serial number A.S.68385 crashing en route to Kelly Field. Two other US Army Air Corps airplanes sustained damage during the filming. (Incidentally, in his autobiography *Squadron of Death*, Richard Grace makes reference to a fatal crash on the *Wings* set: "It was during this period that a certain event happened which had a psychological effect on me. One of the fliers tried to make a turn too quickly and lost his flying speed. He immediately crashed into the ground and was instantly killed. The type of ship he was flying was not much faster than the one I was to crash. The nature of the ground into which he plunged was not as hazardous as hitting posts and shell holes. I began to wonder."

This may, or may not, describe the fate of MB-3A A.S.68385.

It goes without saying that the War Department were very concerned about damage being sustained to government-owned airplanes during the shooting of *Wings*. Nevertheless, given the large amount of pilots and airplanes involved, the safety record obtained during the filming schedule was very good.

The most famous aviation scenes in *Wings* were the intentional crashes of an original SPAD (which would force land into no-man's land and turn turtle) and a Fokker D.VII being 'shot down' while taking off. Both of these crashes were performed by Hollywood's premier 'crash engineer' Richard Grace. In the Fokker D.VII crash, Grace broke his neck, earning him a long hospital stay.

The story of *Wings* begins in early 1926, when Hollywood writer, and former pursuit pilot, John Monk Saunders presented the idea of a movie to be based on aerial combat during Great War to Paramount Pictures producer Jesse L. Lasky. According to Saunders, the air war was a "virgin-province", and "a subject whose proper medium of presentation was the screen." Initially, Lasky balked at Saunders suggestion, stating that the cost would be "appalling", and could such a movie be made on a limited budget? Saunders response was that "It couldn't...If it were attempted at all, it must be done on a grand scale. The very magnitude of the subject demands heroic treatment."

After some deliberation, Paramount agreed to Saunders proposal. The working title for the movie was *Wings*, with Saunders to provide a script outline. Twenty-eight year-old William Wellman, himself a former Lafayette Flying Corps pilot, was chosen to direct the epic. The story involves two American pilots of the 39th Aero Squadron, Jack 'Speed' Powell and David Armstrong (played by Charles 'Buddy' Rogers and Richard Arlen respectively). The leading role as volunteer ambulance driver Mary Preston went to Clara Bow (Hollywood's 'It Girl', who at the time was the world's biggest female movie star). The second main female role of Sylvia Lewis, the love interest of both Powell and Armstrong, went to Jobyna Ralston.

Richard Arlen held one advantage over his co-stars; he could fly. In 1917, Arlen enrolled in the Royal Flying Corps, learning to fly at Camp Borden in Canada. Discharged in early 1919, Arlen made his way to Hollywood the following year, eventually becoming a contract actor. Both he and co-star Rogers were relatively unknown actors when chosen to appear in *Wings*. During the shooting, Rogers had to learn to fly. Although neither Arlen nor Rogers were allowed to fly any of the numerous military airplanes, they nevertheless had to be filmed close-up while in the air. Richard Arlen: "We were using the first motor-driven cameras which were mounted in a lithe in front of the cockpit... The four hundred feet reels ran off at about ninety feet a minute which gave us only a little more than four minutes of picture.

"Bill Wellman would tell us on the ground what he wanted us to do in the air...We would waggle our wings when ready and then take over as producer, director, and actor. We would hold up the proper number of fingers for takes one, two, or three. If we thought the scene was bad we would run a finger across our throats for a cut.

"That was why Wellman wanted actors for *Wings* who could fly. Buddy couldn't. But he learned damn quick!"

(During the Second World War, 'Buddy' Rogers served as a US Navy test pilot.)

Directed by 28-year old William Wellman, a Lafayette Flying Corps veteran, *Wings* was based on the Battle of Saint Mihiel. Recreating the battlefield at Camp Stanley, including trenches, shell craters, and barbed wire involved 600 Mexican workers, who completed the work in two months under the supervision of 2nd Infantry Division engineers. The budget was $ 2 million ($ 28.88 million in 2019 dollars), a huge sum. Among the Air Service pilots flying before the cameras were Hoyt S. Vandenberg and 'Hap' Arnold. Although set in 1918, Wellman had no qualms about using mid-1920s equipment, from uniforms, vehicles, weapons to airplanes. Nevertheless, as described above, the MB-3As were modified to look like SPAD pursuits. Filming was centred at and around Kelly Field outside San Antonio, Texas.

The story in *Wings* revolves around two young pursuit pilots, Jack 'Speed' Powell (Charles 'Buddy' Rogers) and David Armstrong (Richard Arlen). Both are in love with the same girl, Sylvia Lewis (Jobyna Ralston), but she only has feelings for Armstrong. (Incidentally, Ralston would later marry Arlen in real life.) Unbeknownst to Jack, another girl, Mary Preston (Clara Bow) has a crush for him.

Both Powell and Armstrong quickly become experienced pursuit pilots. On one mission, Powell is shot down into no-man's land, but manages to return to his unit. (This scene saw the first of Richard Grace's two crashes for *Wings*.)

While in Paris on leave, Jack gets roaring drunk, ending up with Mary Preston. In the morning, Mary leaves Jack at the hotel, never mentioning her name. He later discovers that his love for Sylvia Lewis is unanswered. Disgusted, he turns against Armstrong, refusing to have anything to do with him.

One day, they are ordered to fly a mission together to shoot down observation balloons. When intercepted by German pursuits, Powell leaves Armstrong to his fate. Disgusted by his own behavior, Powell makes a solitary flight across the enemy lines. Seeing German fighters taking off, he shoots down a Fokker D.VII (Grace's second intentional crash) before dogfighting a second German plane. Powell also manages to shoot this one down, which crashes into a French farm. Landing in a nearby field near the crashed, black-painted MB-3A (thinly-disguised as a Fokker D.VII), Jack is devastated to find a mortally wounded Armstrong in the pilot's seat.

Following the end of the war, Powell returns home. There he meets Sylvia, forcing himself to say that her wait for Armstrong is over, also recounting his role in his friend's death. Powell also get re-acquainted with Mary, whom he does not recognize as the girl he spent the night with in Paris. She does not reveal herself to Jack, either.

Following six months of planning, actual filming began on September 7, 1926, ending exactly seven months later, on April 7, 1927. Recreating the Saint Mihiel battle involved 165 airplanes, 3,500 troops, and many pre-set explosions. Just this scene took a $300,000 slice out of the $ 2 million budget. The aviation sequences took six months to complete. One reason for this was William Wellman's insistence to film the aerial combat sequences against a cloudy sky, declaring that to the general public, viewing airplanes against a blue sky made them seem immovable. The aerial combat scenes were color-tinted in blue, red or yellow to enhance the emotional effect, each frame being hand-painted! A very basic soundtrack was initially used for the noise of the machine guns firing or engines purring away. This was achieved in quite a novel way. Large vertical holes were cut into the film, millimetres away from the sprocket holes. When the film was shown, a lever engaged these holes, creating the noise of machine guns or engines. This required a huge, complex machine, similar to the mechanized nickelodeon popular during the early 20th Century, which out of necessity had to have a purpose-designed railway car for transportation between selected venues. When shown at regular cinemas, *Wings* was a silent movie.

*Wings* premiered at the Criterion Theatre in New York on August 12, 1927, with the west coast premiere, in Los Angeles, taking place on January 15, 1928. With Charles Lindbergh having flown nonstop across the Atlantic only a few months earlier, public interest in aviation was unprecedented. *Wings* was a huge success, constantly breaking attendance records all over America. In May 1929, *Wings* also secured the first Academy Award for Best Picture. Wings eventually grossed $3.6 million at the box office. According to aviation movie historian James H. Farmer, "With *Wings* the airplane had finally come into its own on the screen. In *Wings* it had also forever lost its innocence."

*Wings* footage later appeared in other Paramount movies; *Now We're in the Air* (1927), *Legion of the Condemned* (1928) (which sadly has been lost to history), *Young Eagles* (1930) and *Eagle and the Hawk* (1933).

In 2012, *Wings* was made available on DVD. Along with *Hell's Angels*, released in 1930, *Wings* can be regarded as among the best of all Great War aerial combat movies.

## Thomas-Morse MB-3A Technical Description

The following technical description is quoted from the contemporary Technical Bulletin No. 34

"The standard Boeing MB-3A is an Air Service Type I, single seater pursuit airplane of externally braced biplane construction. It is powered by a standard Wright Model H-3, 320 h.p. engine swinging either a two or four-bladed propeller. The structural features are briefly described in the following:

Wings. – The two-bay cellule of the standard MB-3A has no stagger. It is an externally braced structure with streamline brace wires and spruce interplane struts. The upper wing which carries the gravity fuel tank and the water expansion tank is continuous with two ailerons inserted in the trailing edge at the tips. The lower wing has two panels, which are attached directly to the fuselage wing butt on each side. Wood construction is used thruout.

Fuselage. – The structure is of the conventional wood construction with laminated ash and spruce longerons and wooden cross members braced by diagonal wires with turnbuckle adjustment. The lower wing butt is built integral with the bottom of the fuselage which has a maximum width of 3½ feet across this portion, the width of the fuselage proper being only 2½ feet with a maximum depth of about 4 feet. Fabric covering is used on both wings and fuselage.

Tail Surfaces. – The new tail surfaces shown herewith are of wood construction with fabric covering. The elevator is built about a box spar formed of two "I" section members with flanges joined by means of plywood. Conventional "I" section ribs made with spruce capstrips and mahogany webs in which several lightening holes are cut, are built on to the spars with the capstrips serving as spacers.

The same construction is used for the stabilizer, except that this structure employs two spars, front and rear, which are not joined with plywood at the top and bottom as in the case of the elevator. This plywood, however, extends from the front spar forward to the leading edge on both top and bottom. Lightening holes are cut in the rib webs and also in the plywood along the leading edge.

The rudder and fin are of wood construction also, the former consisting of one main box and a channel section, joined by ribs of the same design as those used in the elevator and by plywood glued to top and bottom flanges. The fin is constructed with a single laminated main spar, to the upper part of which the rudder is hinged. And two small auxiliary vertical spars of plywood to which the ribs are joined. The leading edge is made of laminated spruce milled and routed to shape. The whole structure is covered with 3-ply mahogany plywood with large lightening holes cut thru between the spars and ribs.

The general construction of the old style tail surfaces with the short vertical fin and balanced elevator and rudder is somewhat similar to that used in the new tail surfaces. These are used on the first one hundred fifty airplanes built under this contract.

Chassis. – Spruce struts supporting a continuous streamline axle fitted with elastic cord shock absorber and 28 by 4 inch straight-side wheels and Goodyear tires comprise the two-wheeled landing chassis used on this airplane. The axle, proper, consists of the two short shafts, one for each wheel, supported between the spreader tubes. The structure is strengthened by vertical streamline brace wires.

Power Plant. – The power plant consists of a standard Wright H-3 engine capable of developing 320 horse power at a normal speed of 1800 r.p.m. The engine is fitted with one Stromberg NAD-6A carbutetor with standard mixture control effective to ceiling, and an automatic spark advance, ignition being furnished by two Dixie magnetos.

Fuel system comprises two leak-proof tanks, the main tank of 33 gallons capacity being located in bottom of fuselage directly in front of pilot and the 12-gallon gravity tank being placed in the center section of upper wing. Fuel is lifted from main tank by suction thru a line strainer to an engine-driven bellows pump which forces it thru a 3-way cock directly to carburettor or to both carburettor and gravity tank, the latter overflowing into the main tank. Gravity pressure alone may be used to operate system is case of pump failure.

Water cooling is effected by means of two side radiators of Thomas-Morse manufacture, supported on padded ash sills on each side of fuselage opposite cockpit. The expansion or reserve tank, which is provided with an automatic pressure relief valve to prevent excessive loss of water under adverse conditions, is placed in the upper wing. Part of the water which has been heated by the engine, is circulated around the intake manifold, thus heating the gasoline mixture. Cooling is regulated by shutters over radiators. These are controlled from the cockpit.

Equipment. – In addition to the requisite flying equipment, this airplane carries oxygen and radio apparatus, two life preserver cushions and fire extinguisher. Provision for carrying camera is made also. Armament consists of two .32 caliber synchronized aircraft guns with necessary ammunition boxes, chutes, and sights. One of the airplanes at McCook Field has been equipped with electric landing lights and other night flying accessories."

## MB-3A Technical Data and Performance Characteristics

Type: Single-seat pursuit/advanced trainer
Engine: 1 x 320 h.p. Wright H-3
Wingspan: 26 ft 0 in (upper) (7,92 m) 24 ft 6 in (lower)
Length: 20 ft (6,0958 m)
Height: 7 ft 8 in (2,59 m)
Wing area: 228 sq ft (21,23 m$^2$)
Empty weight: 1,716 lb (779 kg)
Military load: 767,82 lb (348,59 kg)
Armament: 205,32 lb (93,21 kg)
Equipment: 84,5 lb (38,36 kg)
Fuel: 264 lb (119,85 kg)
Oil: 34 lb (15,43 kg)
Empty weight: 1,716 lb (778 kg)
Gross weight: 2,485 lb (1128,2 kg)
Maximum speed at sea level: 140 mph (225 km/h)
Cruise speed: 125 mph (201 km/h)
Climb rate: 1,350 ft./min (411 m)
Service ceiling: 19,500 ft (5,943 m)
Endurance: 2.25 hrs
Armament: 2 x 0.32 machine guns

## Colors and Markings

The four MB-3 prototypes were finished in the standard 1918 camouflage; olive drab (or khaki brown) on the upper surfaces and vertical fin, with all other areas being cream (or ivory). The latter included the wing struts, etc. The pair of MB-3 prototypes which were transferred to McCook Field, 40093 and 40095, had been repainted overall olive drab by mid-1920. At least one, and most likely all four prototypes, were marked with the red, blue, and white roundel (official national insignia between January 1918 and August 1919) with the rudder being adorned with blue, white, and red (forward) vertical stripes. The respective serial number, 40092 to 40095 appeared prominently in large black figures on both sides of the fuselage without the A.S. prefix.

In August 1919, the national insignia reverted to the old five-pointed star with a red circle superimposed on a blue roundel. This was carried in four positions, on the upper and lower surfaces of the wings. The MB-3s, as well as the MB-3As were finished in overall olive drab. The respective serial number was applied in large numerals, usually along with the A.S. prefix. The size of the serial numbers were later reduced in size to make room for unit insignia. Airplanes within a particular unit usually carried an individual numeral, which was painted on the fuselage. Occasionally, this was repeated on the wings. The MB-3s which participated in John J Mitchell Trophy Race were adorned with individual race numerals in white on both sides of the fuselage.

The MB-3s used by the US Marine Corps were finished overall olive drab, with national insignia in four positions. The engine cowling was painted grey. Additionally, the USMC insignia appeared in dark red on both sides of the fuselage. The respective serial numbers, eg A-6065, appeared vertically in black on both sides of the rudder. An individual serial number also appeared in white on both sides of the fuselage. Two diagonally stripes (possibly in white) were painted on the upper surfaces on the wings.

The MB-3 and MB-3As tested at McCook Field retained their Air Service colors. A McCook Field serial number, eg P274, was applied on both sides of the rudder.

The MB-3As which appeared in *Wings* were camouflaged in French and German colors à la 1918.

## Appendix I: Production List

### MB-3

Four prototypes

40092

**40093** To McCook Field P66 Parts used to rebuild P121

**40094** To McCook Field P124 Shipped to Thomas-Morse in 1920, Pulman

**40095** To McCook Field P121

First production contract
Fifty aircraft:

63331 to 63380

63331

**63332** To McCook Field P164 To Boeing on June 16, 1921

63333

63334

**63335** To NACA Langley Memorial Aeronautical Laboratory January 1 – November 1, 1923 as NACA 8. Returned to USAAS at unknown date

**63336** To McCook Field P197 Salvaged August 17, 1923

63337

63338

63339

63340

**63341** W/o Ithaca March 30, 1921

**63342** Modified as MB-3M

**63343** Landing accident due to mechanical failure at Langley Field, VA, August 21, 1921. Pilot Byrne V Baucom

63344

63345

63346

63347

63348

**63349** Ground looped and overturned on landing Ellington Field, TX, May 12, 1922. Pilot Hugh M Elmendorf. Also listed as crashed in landing accident at Ellington Field on May 13, 1922. Pilot Charles H Howard

63350

63351

63352

63353

63354

63355

63356

**63357** Ground looped on landing Ellington Field, TX, May 13, 1922. Pilot Benjamin R McBride

63358

63359

63360

63361

63362

63363

63364

63365

63366
63367
63368
63369
63370
63371
63372
63373
63374
63375
63376
63377
63378
63379
63380
A.S.64374 to A.S.64384 Originally built to an Air Service contract, but transferred to USMC as A-6060 to A-6070. Survivors sold back to the Air Service in 1923.

## MB-3A

200 built by Boeing

68237 to 68436 c/ns 210 to 409

68237 c/n 210
**68238** c/n 211 McCook Field P260 Wrecked in April 1923
68239 c/n 212
68240 c/n 213
68241 c/n 214
68242 c/n 215
68243 c/n 216
**68244** c/n 217 ASTTS Based Chanute Field, IL. Lost, cause unknown, on April 17, 1923. Pilot Louis N Eller
68245 c/n 218
68246 c/n 219
**68247** c/n 220 95th PS Overturned in take-off accident at Bridgetown Field, St Louis, MO, September 23, 1923. Pilot Thomas K Mathews
68248 c/n 221
**68249** c/n 222 94 PS Overturned on landing at Selfridge Field, MI, March 12, 1924
68250 c/n 223
**68251** c/n 224 Overturned on landing at air mail field, Bryan, OH, July 13, 1923. Pilot George P Tourtellot
68252 c/n 225
68253 c/n 226
**68254** c/n 227 Force landing due to engine failure, presumed near Selfridge Field, MI, January 25, 1923. Pilot Laclair D Schultz
68255 c/n 228
68256 c/n 229
68257 c/n 230
**68258** c/n 231 Force landed due to engine failure at Chaffee, NY, November 11, 1923. Pilot Frank O'D Hunter
**68259** c/n 232 95th PS Force landed due to engine failure 11 mi NE Marshall, MI, August 25, 1923. Pilot L M Abby
68260 c/n 233
**68261** c/n 234 27th PS # 54 Wrecked in January 1923
68262 c/n 235
68263 c/n 236
**68264** c/n 237 18th HS General Billy Mitchell's personal MB-3A. Named 'Hawk'. Accident in December 1924
**68265** c/n 238 27th PS Wrecked in June 1924
**68266** c/n 239 27th PS Overturned on landing at Selfridge Field, MI, 2Oct 1922. Pilot Burt E Skeel
68267 c/n 240
68268 c/n 241
**68269** c/n 242 Surveyed in March 1927
**68270** c/n 243 Surveyed in August 1926
68271 c/n 244
**68272** c/n 245 Wrecked in May 1925
**68273** c/n 246 27th PS Overturned during forced landing due to engine failure 1½ mi SW Decatur, MI, November 27, 1923
68274 c/n 247
**68275** c/n 248 Surveyed in August 1926
68276 c/n 249
68277 c/n 250
**68278** c/n 251 19th PS Wrecked in Hawaii in April 1926
68279 c/n 252
**68280** c/n 253 95th PS Wrecked in March 1924
**68281** c/n 254 To McCook Field as P353. Accident in October 1924
**68282** c/n 255 W/o in gunnery accident at Selfridge Field, MI, March 18, 1924. Pilot Reuben C Moffat
**68283** c/n 256 Surveyed in October 1926
68284 c/n 257
**68285** c/n 258 Overturned during forced landing due to engine failure 3 mi NE Buchanan, MI, December 1, 1923
68286 c/n 259
**68287** c/n 260 Surveyed in August 1927
**68288** c/n 261 Surveyed in August 1926
68289 c/n 262
**68290** c/n 263 Headquarters Flight, 1st PG Wrecked in October 1923
**68291** c/n 264 Camp Lewis Wrecked in September 1922
**68292** c/n 265 Surveyed in August 1927
68293 c/n 266
68294 c/n 267

**68295** c/n 268 17th PS? Accident in December 1923
68296 c/n 269
**68297** c/n 270 19th PS # 29 Wrecked in July 1925
68298 c/n 271
68299 c/n 272
**68300** c/n 273 Surveyed in January 1927
68301 c/n 274
**68302** c/n 275 Surveyed in August 1927
68303 c/n 276
**68304** c/n 277 Surveyed in September 1926
68305 c/n 278
**68306** c/n 279 17th PS Force landed due to engine failure NW Detroit, MI, October 26, 1923. Pilot Oliver W Broberg
**68307** c/n 280 Surveyed in November 1926
68308 c/n 281
68309 c/n 282
**68310** c/n 283 17th PS # 11 Accident in April 1924
**68311** c/n 284 Surveyed in December 1926
68312 c/n 285
68313 c/n 286
**68314** c/n 287 Surveyed in January 1927
**68315** c/n 288 Landing accident at Selfridge Field, MI, July 27, 1923. Pilot Russell L Meredith
68316 c/n 289
**68317** c/n 290 Surveyed in November 1926
68318 c/n 291
68319 c/n 292
**68320** c/n 293 ASTS Wrecked in October 1925
68321 c/n 294
**68322** c/n 295 19th PS Wrecked in Hawaii in November 1925
68323 c/n 296
68324 c/n 297
68325 c/n 298
68326 c/n 299
68327 c/n 300
68328 c/n 301
**68329** c/n 302 Selfridge Field Wrecked in April 1925
68330 c/n 303
68331 c/n 304
**68332** c/n 305 Surveyed in September 1926
68333 c/n 306
68334 c/n 307
**68335** c/n 308 Surveyed in September 1926
68336 c/n 309
**68337** c/n 310 Surveyed in September 1926
68338 c/n 311
68339 c/n 312
**68340** c/n 313 Surveyed in June 1927
68341 c/n 314
68342 c/n 315
68343 c/n 316
**68344** c/n 317 24th PS Accident in the Canal Zone in May 1925
68345 c/n 318
68346 c/n 319
68347 c/n 320
68348 c/n 321
**68349** c/n 322 24th PS Wrecked in the Canal Zone in January 1927
**68350** c/n 323 Surveyed in April 1927
68351 c/n 324
**68352** c/n 325 Surveyed in November 1926
68353 c/n 326
**68354** c/n 327 43rd SS # 21 Accident in June 1925
68355 c/n 328
68356 c/n 329
68357 c/n 330
68358 c/n 331
68359 c/n 332
68360 c/n 333
**68361** c/n 334 Surveyed in April 1927
68362 c/n 335
**68363** c/n 336 Surveyed in November 1926
**68364** c/n 337 24th PS # 1 Wrecked in the Canal Zone in August 1925
**68365** c/n 338 Surveyed in January 1927
68366 c/n 339
68367 c/n 340
**68368** c/n 341 Surveyed in December 1926
**68369** c/n 342 Surveyed in January 1927
**68370** c/n 343 19th PS Wrecked in Hawaii in April 1924
**68371** c/n 344 19th PS Wrecked in Hawaii in July 1925
**68372** c/n 345 43rd SS Flew in 'Wings' as # 7
**68373** c/n 346 19th PS Landing accident due to engine failure, Wheeler Field, HI, February 15, 1924. Pilot Floyd A Lundell
**68374** c/n 347 19th PS Wrecked in Hawaii in April 1924
68375 c/n 348
**68376** c/n 349 Surveyed December 1926
68377 c/n 350
**68378** c/n 351 6th PS Wrecked in Hawaii, July 1924
**68379** c/n 352 6 PS # 9 Cat. 5 accident at Luke Field, Oahu, HI, November 6, 1924. Pilot Milo N. Clark Also reported as surveyed in December 1929
**68380** c/n 353 Wrecked at Camp Lewis, November 1922
**68381** c/n 354 19th PS # 25 Accident in September 1924
**68382** c/n 355 Surveyed in August 1926
**68383** c/n 356 Surveyed in January 1927
**68384** c/n 357 Surveyed in January 1927
**68385** c/n 358 43rd SS Crashed en route to Kelly Field in November 1926 in connection with filming of Wings Painted as # 3

**68386** c/n 359 Surveyed in August 1926
68387 c/n 360
**68388** c/n 361 19th PS # 7 Accident in Hawaii in April 1924
**68389** c/n 362 Take-off accident due to engine failure at Luke Field, Ford Island, Oahu, HI, 11Mar 1924. Pilot Harold A Moore
**68390** c/n 363 6th PS # 5 Accident in May 1924
**68391** c/n 364 6th PS # 5 Surveyed in December 1926
68392 c/n 365
68393 c/n 366
68394 c/n 367
68395 c/n 368
**68396** c/n 369 Surveyed in September 1926
68397 c/n 370
**68398** c/n 371 Surveyed in January 1927
68399 c/n 372
**68400** c/n 373 Surveyed in January 1927
**68401** c/n 374 3rd PS Accident in the Philippines in March 1926
68402 c/n 375
**68403** c/n 376 Surveyed in January 1927
68404 c/n 377
68405 c/n 378
**68406** c/n 379 3rd PS W/o in take-off accident due to engine failure, 400 yards NE Clark Field, Pampanga, Philippines. Pilot J.C. McDonnall
68407 c/n 380
**68408** c/n 381 3rd PS Accident in the Philippines in December 1924
**68409** c/n 382 Surveyed in January 1927
68410 c/n 383
68411 c/n 384
68412 c/n 385
**68413** c/n 386 Surveyed in January 1927
68414 c/n 387
**68415** c/n 388 Surveyed in January 1927
68416 c/n 389
68417 c/n 390
68418 c/n 391
**68419** c/n 392 43rd SS Wrecked in July 1925
**68420** c/n 393 43rd SS Overturned on landing at Kelly Field, TX, 27Feb 1924. Pilot Auby C Strickland
68421 c/n 394
**68422** c/n 395 43rd SS # 6 Taxiing accident at Kelly Field, TX, 4June 1923. Pilot Duane C Warner
**68423** c/n 396 43rd SS Overturned during forced landing due to engine failure 4 mi N of Kelly Field, TX, 1June 1923. Pilot George O Robertson
68424 c/n 397
68425 c/n 398
68426 c/n 399
**68427** c/n 400 43rd SS # 7 Accident in July 1925
68428 c/n 401

## Appendix II: MB-3 & MB-3A Evaluated at McCook Field

| McCook Serial Number | USAAS Serial Number | Remarks |
|---|---|---|
| P66 | 40093 | Second MB-3 Prototype |
| P121 | 40095 | Fourth MB-3 Prototype. A.S. serial number 13629 "To Museum 216.82" (Sadly, all aircraft which had been placed in the care of this forerunner of the National Museum of the USAF, were scrapped in the late 1920s/early 1930s) |
| P164 | 63332 | Second Series Production MB-3. Used to devise improved engine cooling system |
| P197 | 63336 | Sixth Series Production MB-3 |
| P201 | 63338 | Eighth Series Production MB-3 |
| P208 | 63337 | Seventh Series Production MB-3 |
| P259 | 68237 | First Series Production MB-3A. Retro-fitted with Boeing-designed empennage |
| P260 | 68238 | Second Series Production MB-3A. Modified with 'speed wings', which had a new airfoil and four ailerons |
| P274 | '68586' | MB-3A. Serial number as noted in McCook Field records. Actually either 68286 or 68386. Fitted with Wright H engine serial number 68701. Experimentally fitted with four-blade propeller, skis, and night landing lanterns. Shipped to Fairfield, OH, on September 3, 1925. "ST 18393" |
| P353 | 68281 | MB-3A. Surveyed on December 19, 1924 |
| No less than six MB-3s and four MB-3As were tested at McCook Field. | | |

**68429** c/n 402 43 SS # 8 Accident in February 1925
68430 c/n 403
68431 c/n 404
68432 c/n 405
68433 c/n 406
**68434** c/n 407 Crashed in poor weather near Lexington, TX, July 2, 1923. Pilot Byrne V Baucom
68435 c/n 408
68436 c/n 409

# References

## Printed (books and articles) sources:


Andrade, John M. *U.S. Military Aircraft Designations and Serials 1909 to 1979*. Midland Counties Publications, Leicester, United Kingdom 1997 (Second Printing)

Bliss, R.L. 'The American SPAD'. In *Skyways* No. 11, July 1989

Bowers, Peter M. *Boeing Aircraft since 1916.* Putnam and Company Ltd, London, Great Britain 1966

Bowers, Peter M. 'The First Pursuit The Life and, Generally, Hard Times Of The MB-3!' *Wings*, December 1976 issue

Bowers, Peter M. *Curtiss Aircraft, 1907–1947.* Putnam & Company Ltd, London 1979

Farmer, James H. *Celluloid Wings – The Impact of Movies on Aviation.* TAB Books Inc, Blue Ridge Summit, PA, USA 1984

Brown, K.S., Heyn, E.F., Freeman, R.A., Bowyer, M.J.F. & Berry, P. *United States Army and Air Force Fighters 1916–1961*. Aero Publishers, Inc. Fallbrook, CA, USA 1961

Cavanagh, Robert L & Rust, Kenn C. 'Early Air Service Pursuits, Parts I and II.' In *Skyways*, No. 15 & 16, July and October 1990

Eberspacher, Warren. 'Thomas-Morse M.B.4' in *A.A.H.S. Journal*, Fall 1967

Farmer, James H. *Celluloid Wings The Impact of Movies on Aviation.* TAB Books, Inc. Blue Ridge Summit, PA, USA 1984

Foxworth, Thomas G. 'Thomas Morse Racing Airplanes.' In *A.A.H.S. Journal*, Winter 1967

Gough, Michael. 'Air Races and the Greatest U.S. Air Meet, St. Louis, 1923.' In *A.A.H.S. Journal*, Vol. 59, No. 3/4, Winter 2014

Grace, Richard. *Squadron of Death The True Adventures of A Movie Plane-Crasher.* Doubleday, Doran & Company, Inc. Garden City, New York, USA 1929

Green, William & Swanborough, Gordon. *The Great Book of Fighters.* MBI Publishing Company, Osceola, WI, USA 2001

Grey, C.G. *Jane's All The World's Aircraft*. 1919 edition. Facsimile reprint David & Charles, Newton Abbot, England 1969

Hamady, Theodore. *The Nieuport 28 America's First Fighter.* Schiffer Publishing Ltd, Atglen, PA, USA 2008

Matt, Paul R. *United States Navy and Marine Corps Fighters 1918–1962.* Aero Publishers, Inc. Fallbrook, CA, USA 1962

Norton, F.H. 'Pressure Distribution Over the Wings of an MB-3 Airplane in Flight.' In *Skyways* No. 22, April 1992

Owers, Colin. *Thomas Morse S-4 Scout.* Albatros Productions, Ltd., Berkhamstead, Great Britain 2014

Pearcy, Arthur. *Flying the Frontiers NACA and NASA Experimental Aircraft.* Airlife Publishing Ltd, Shrewsbury, England 1993

Ronnie, Art. 'Box Office Ace.' In *Air Classics*, February 1969

Swanborough, Gordon & Bowers, Peter M. *United States Navy Aircraft since 1911*. Putnam Aeronautical Books, London, Great Britain 1990

Technical Bulletin No. 34

Weaver, Truman C. 'The Thomas-Morse Racer.' In *Air Progress*, Vol 21 Numer 4, October 1967

Wegg, John. *General Dynamics Aircraft and their Predecessors.* Putnam Aeronautical Books, London, England 1990


## Internet:


www.aviationarchaeology.com
http://aviation-safety.net


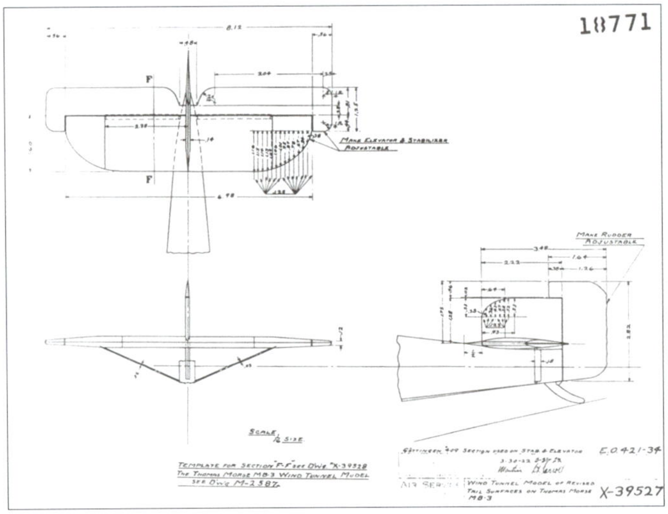


**Left:** A factory drawing showing details of the MB-3 wind tunnel model. Via Colin Owers

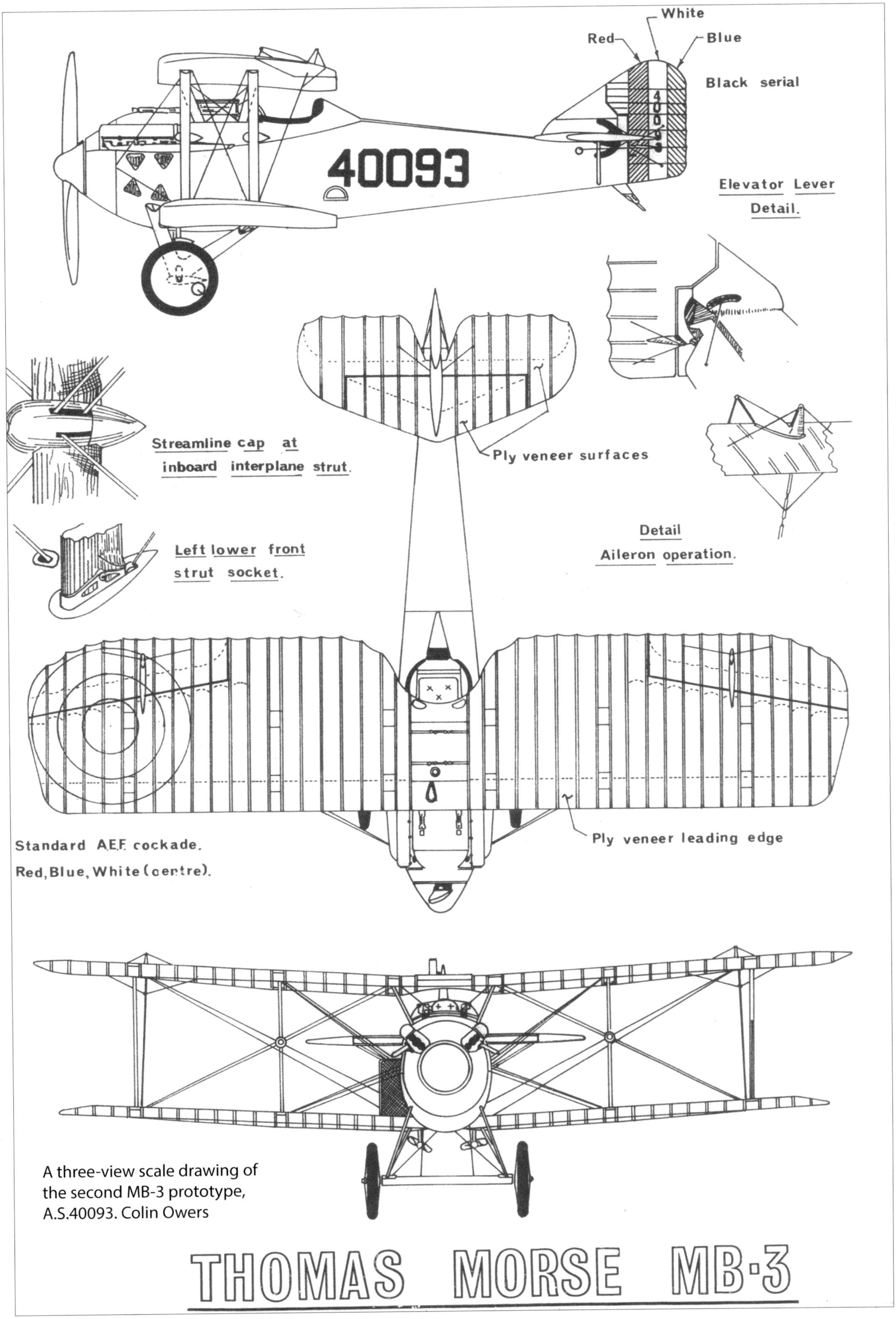


A three-view scale drawing of the second MB-3 prototype, A.S.40093. Colin Owers

The second Thomas-Morse MB-3 prototype, A.S.40093.

Thomas-Morse MB-3 63334, the fourth production MB-3.

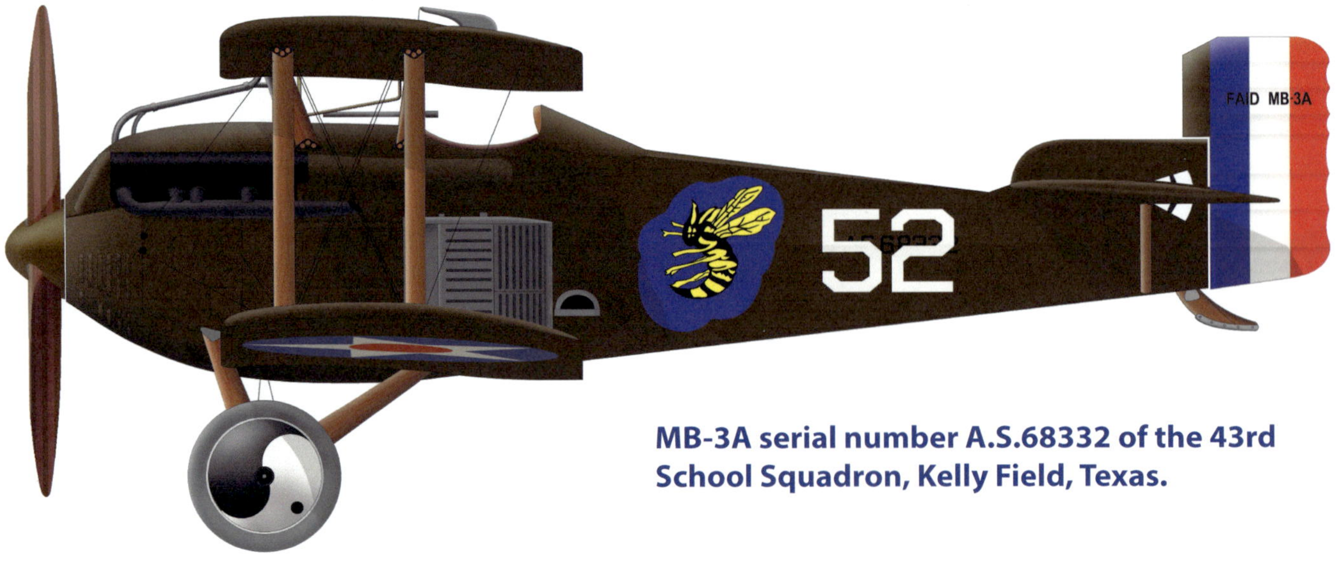


MB-3A serial number A.S.68332 of the 43rd School Squadron, Kelly Field, Texas.

Brigadier General Billy Mitchell had a personal MB-3 at his disposal. This particular airplane was nicknamed 'Hawk', being finished overall jet black.

The first USMC MB-3, serial number A-6060.

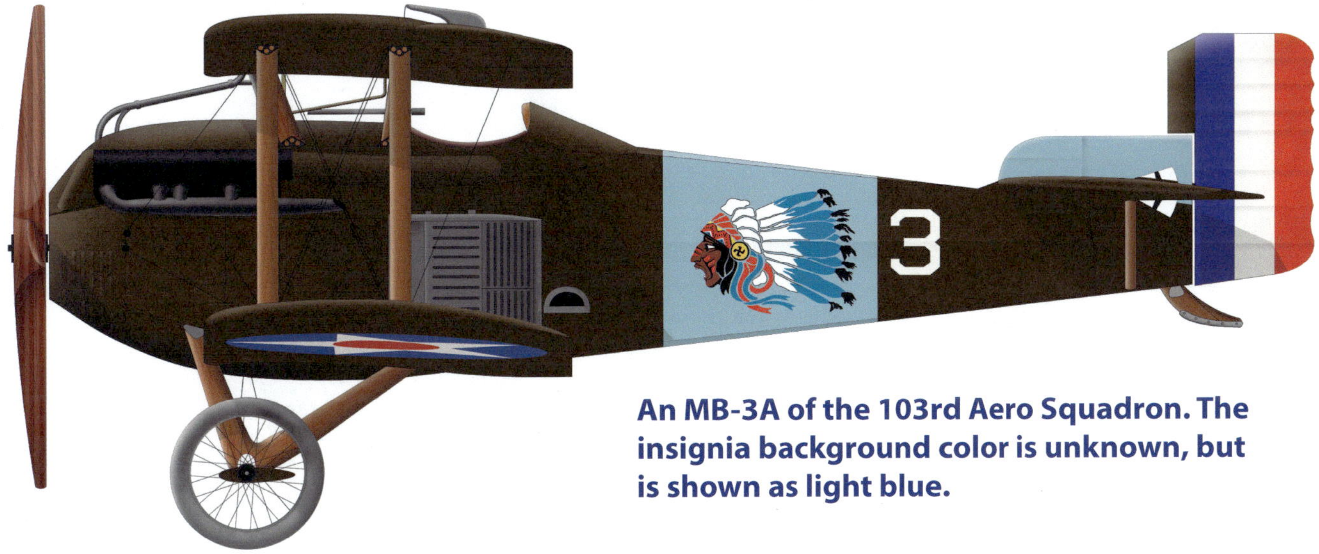


An MB-3A of the 103rd Aero Squadron. The insignia background color is unknown, but is shown as light blue.

Side and upper view of a USMC MB-3.

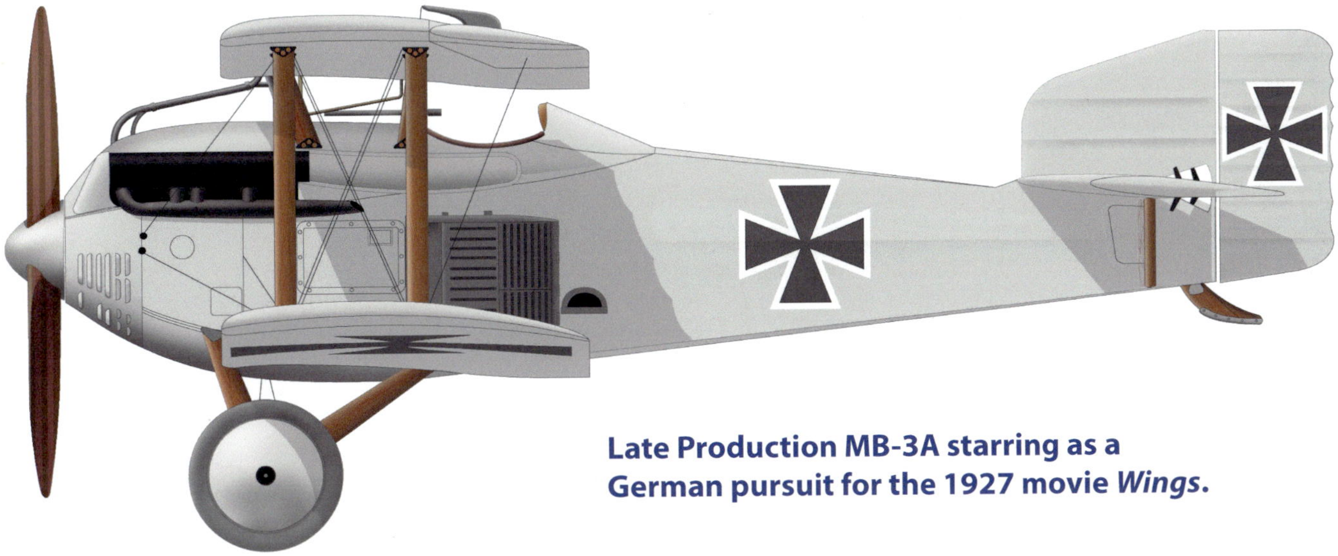

Late Production MB-3A starring as a German pursuit for the 1927 movie *Wings*.

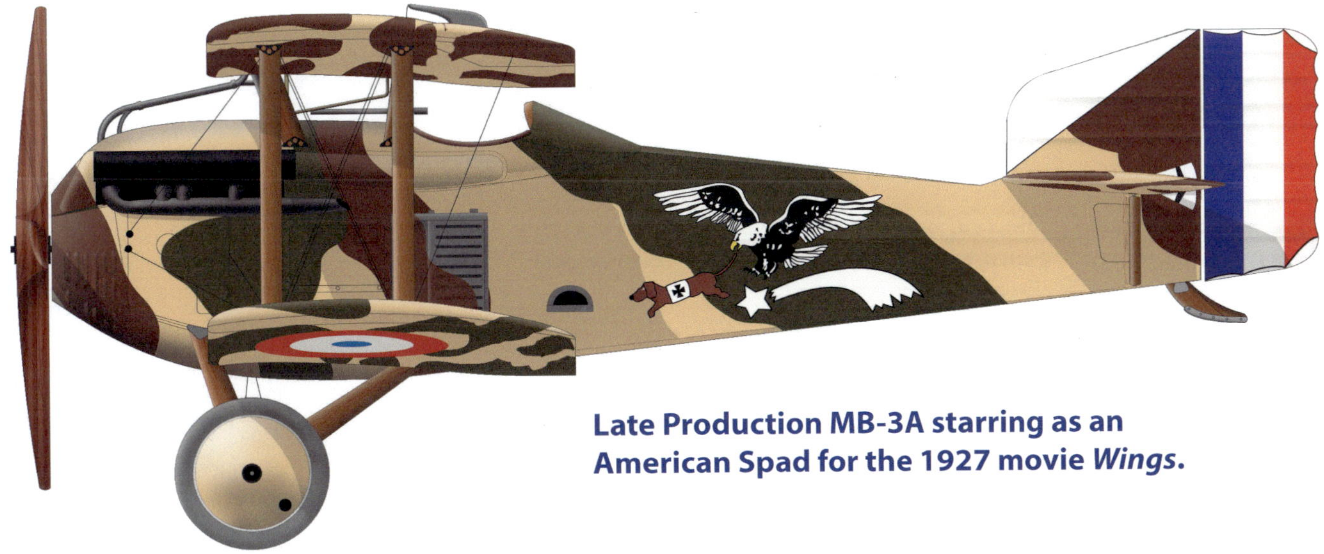

Late Production MB-3A starring as an American Spad for the 1927 movie *Wings*.

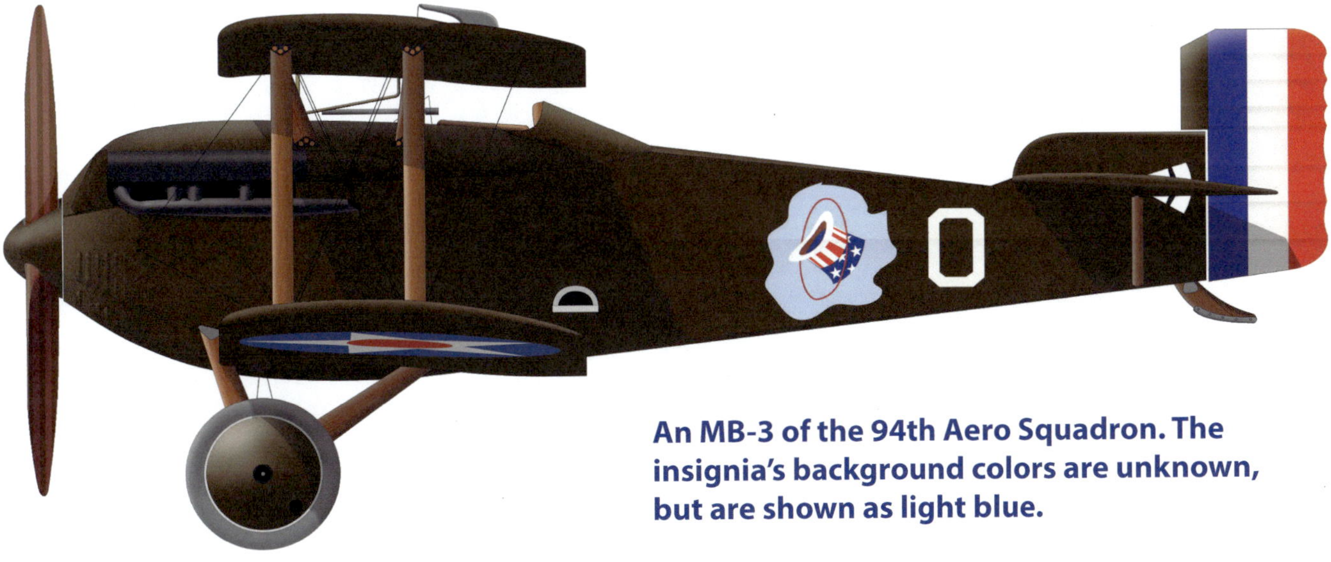


An MB-3 of the 94th Aero Squadron. The insignia's background colors are unknown, but are shown as light blue.

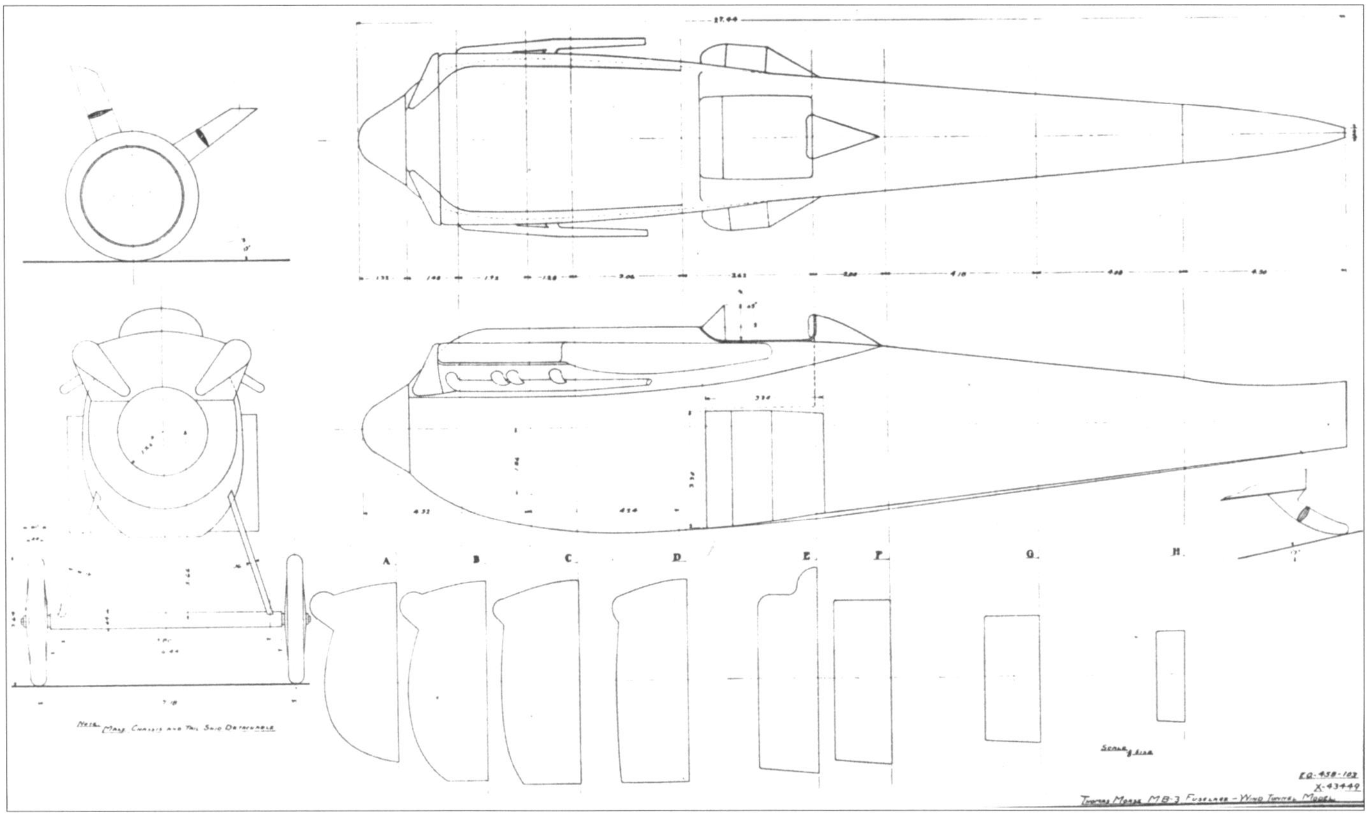

**Above:** Factory drawings showing the fuselage cross sections of the wind tunnel model. Via Colin Owers
**Below:** A factory three-view drawing of the MB-3 wind tunnel model. Via Colin Owers

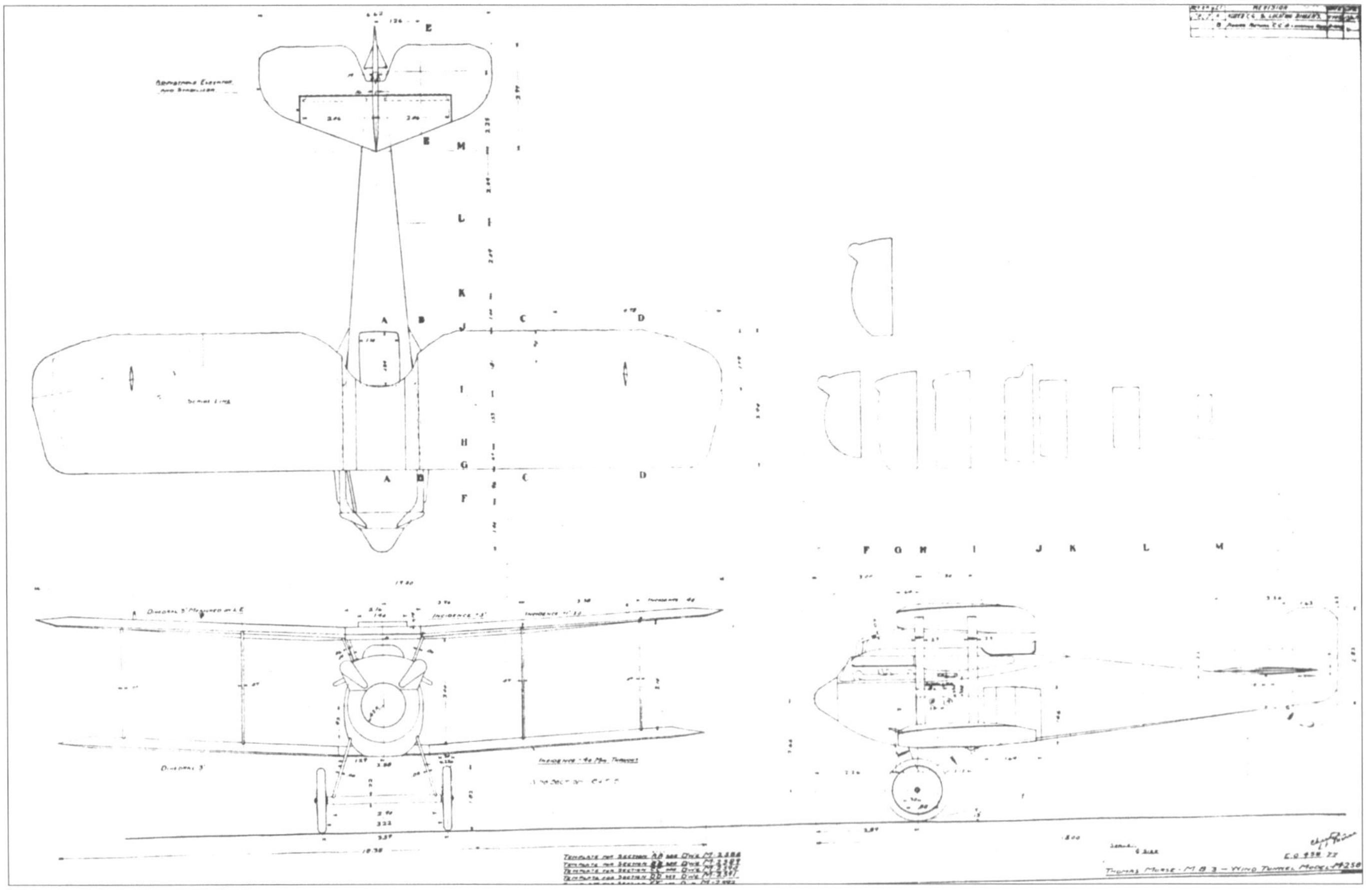

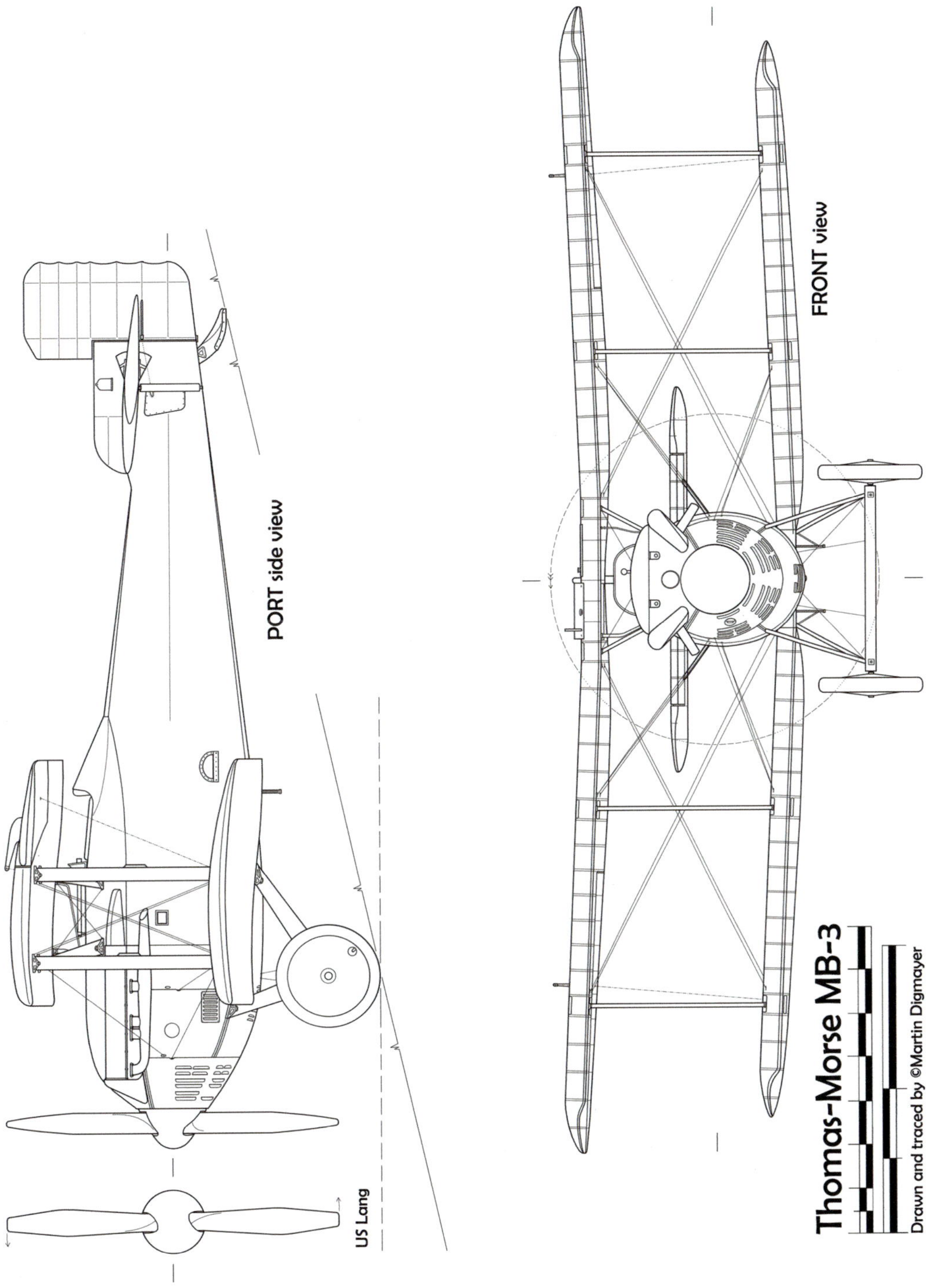
PORT side view
US Lang
FRONT view
Thomas-Morse MB-3
Drawn and traced by ©Martin Digmayer

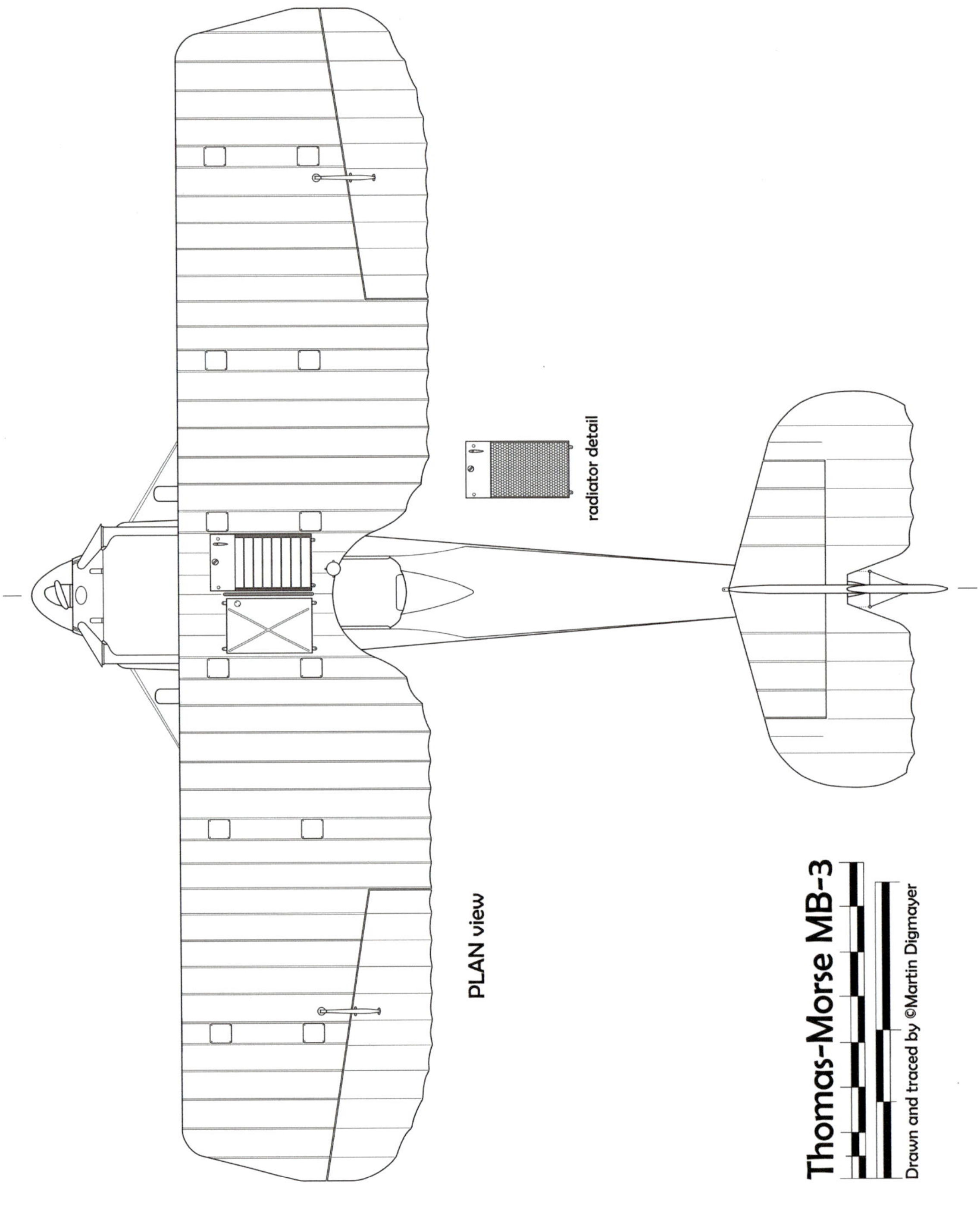
radiator detail
PLAN view
Thomas-Morse MB-3
Drawn and traced by ©Martin Digmayer

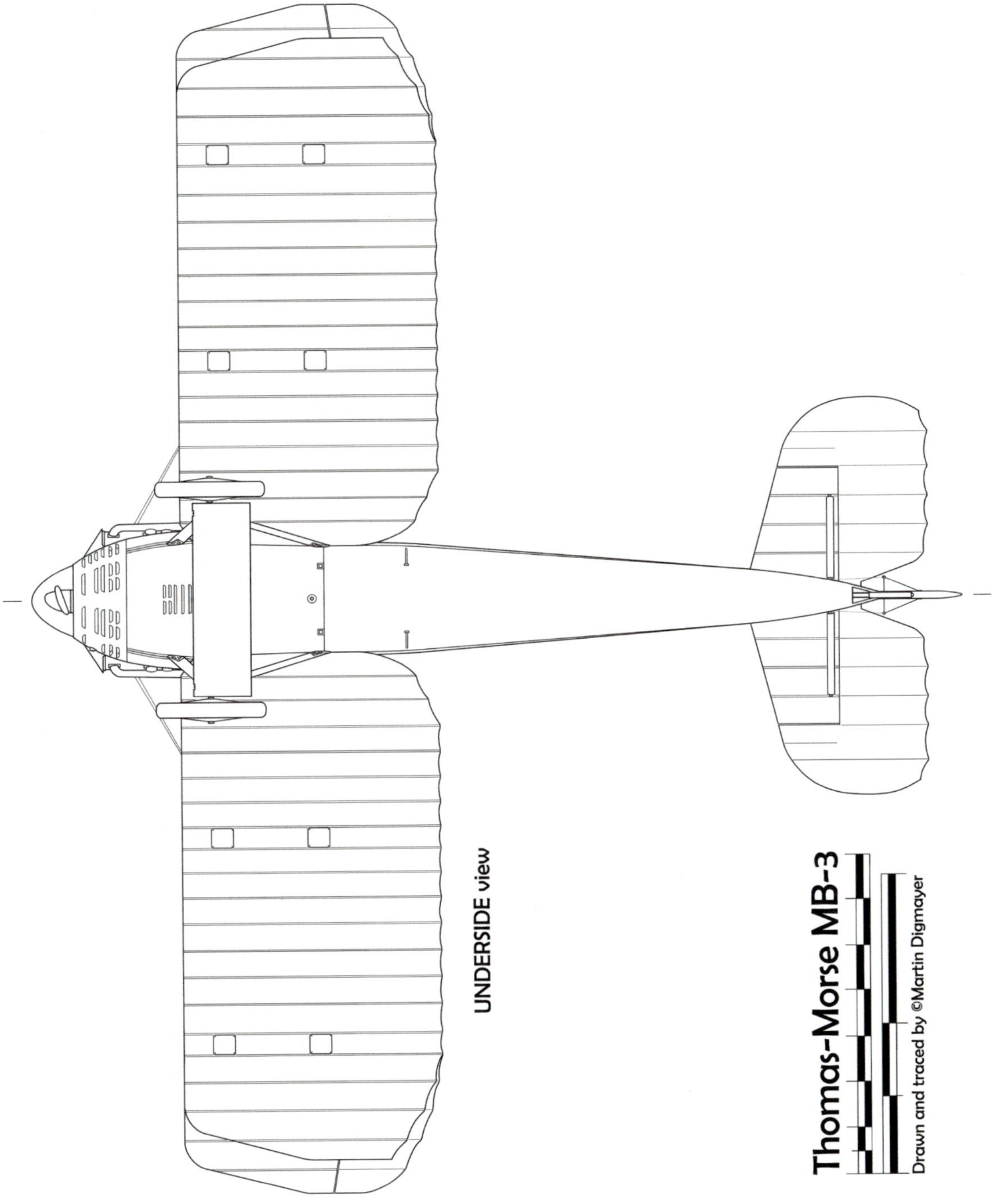
UNDERSIDE view
Thomas-Morse MB-3
Drawn and traced by ©Martin Digmayer

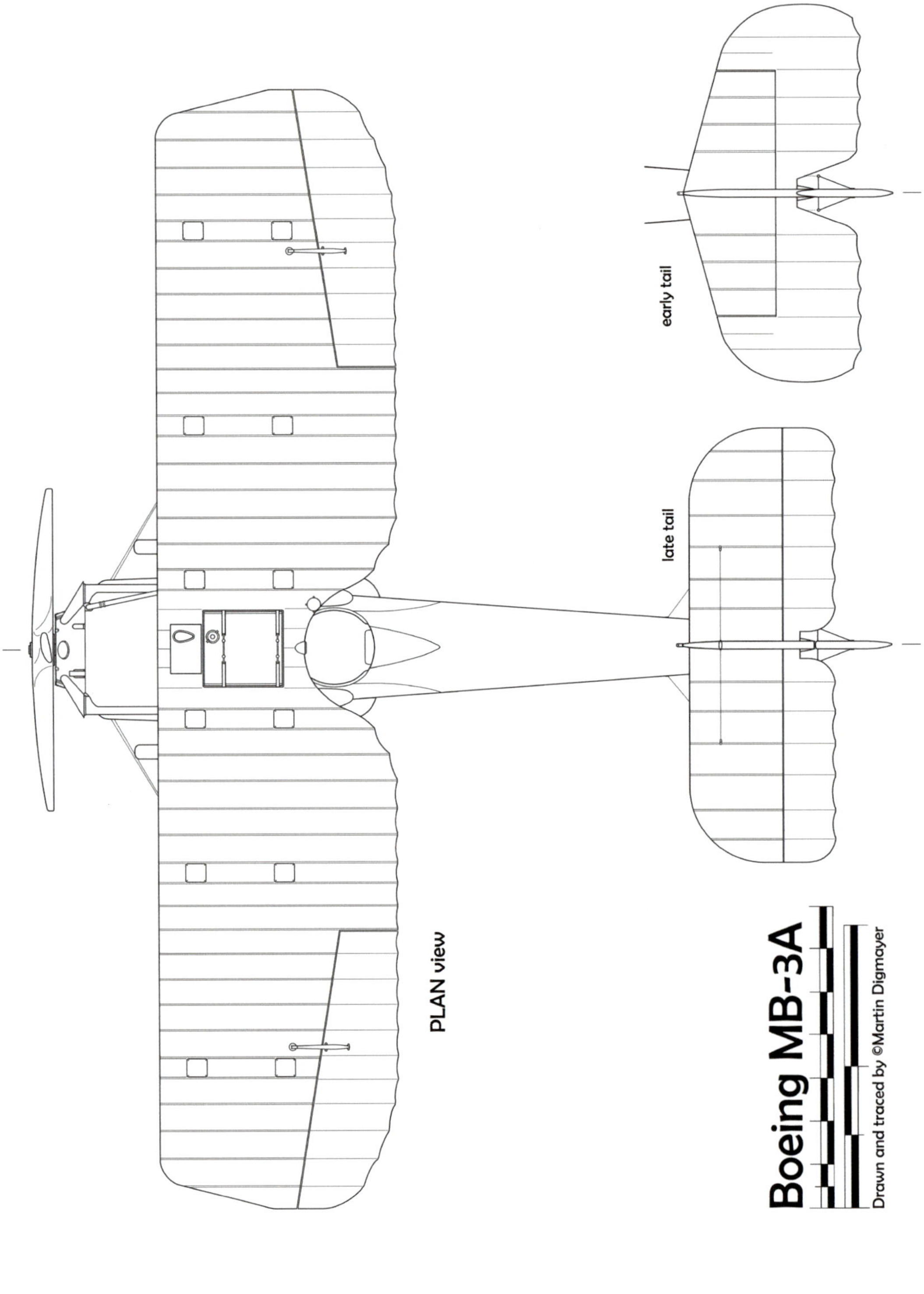
early tail
late tail
PLAN view
Boeing MB-3A
Drawn and traced by ©Martin Digmayer

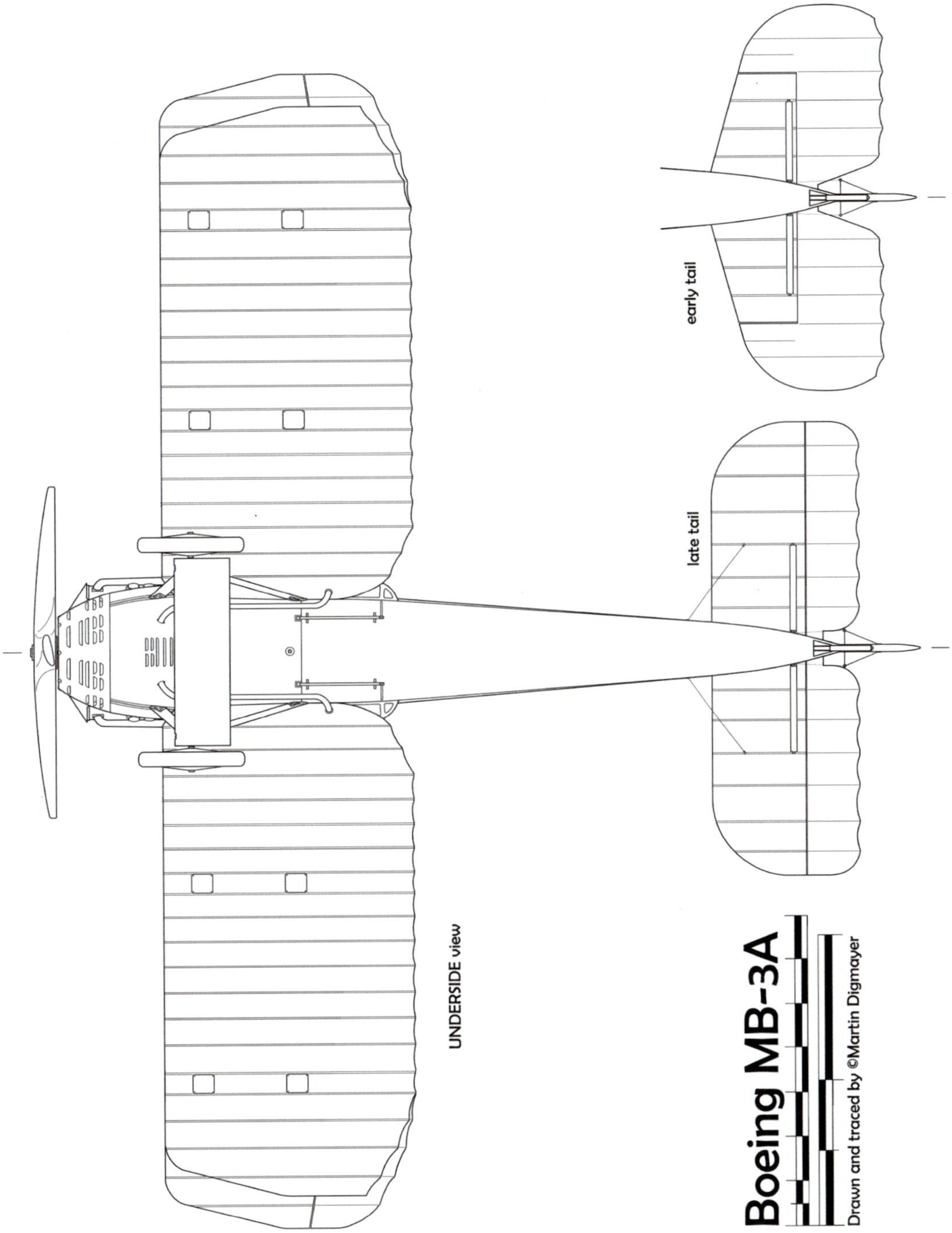
early tail
late tail
UNDERSIDE view
Boeing MB-3A
Drawn and traced by ©Martin Digmayer

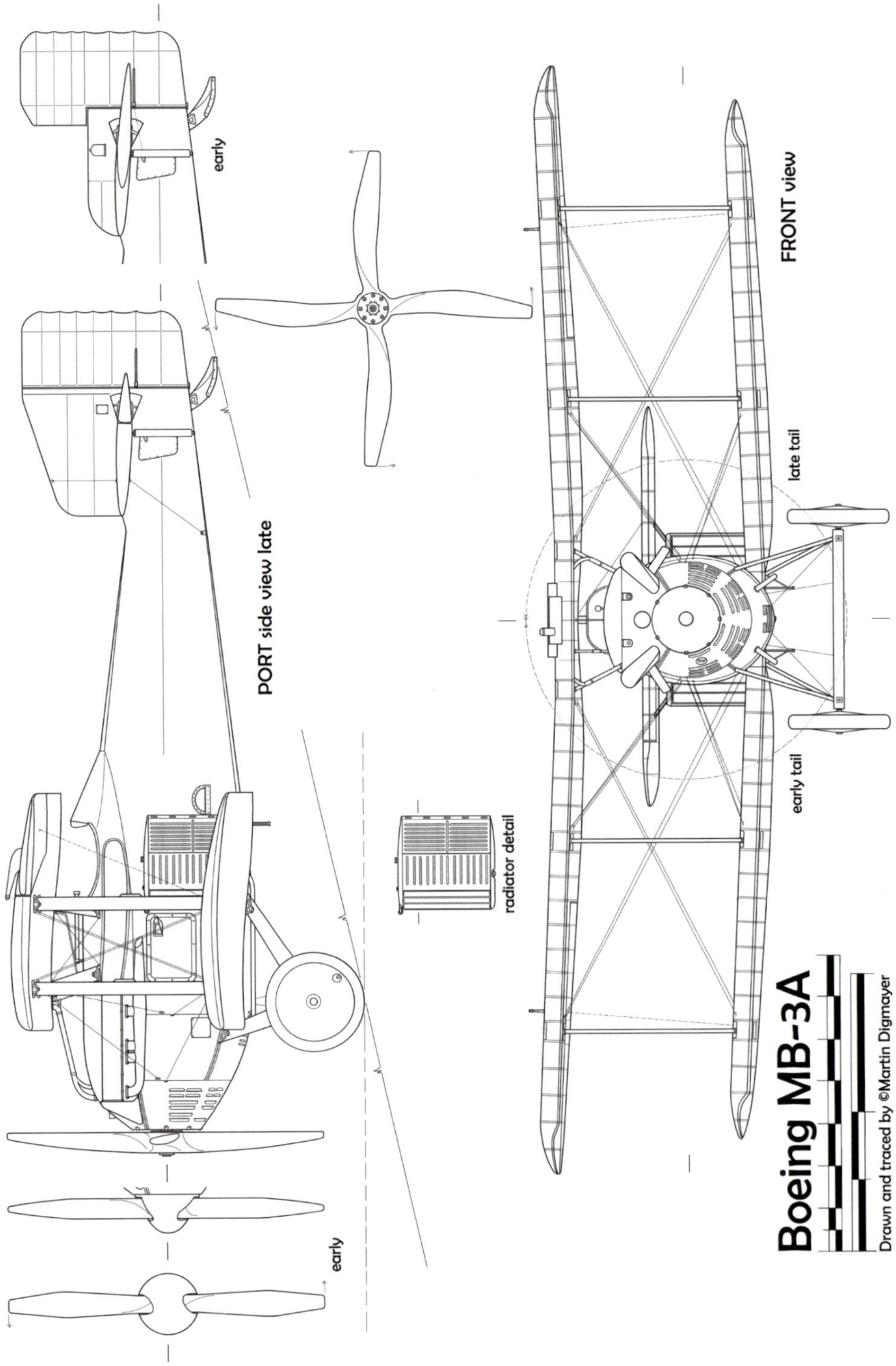
early
PORT side view late
FRONT view
late tail
early tail
radiator detail
Boeing MB-3A
Drawn and traced by ©Martin Digmayer
early

Made in the USA
Las Vegas, NV
10 November 2026

57843572R00052